Ballistic Missile Defense in the Post–Cold War Era

Ballistic Missile Defense in the Post–Cold War Era

David B.H. Denoon

Routledge
Taylor & Francis Group

NEW YORK AND LONDON

First published 1995 by Westview Press, Inc.

Published 2021 by Routledge
605 Third Avenue, New York, NY 10017
2 Park Square, Milton Park, Abingdon, Oxon OX14 4RN

Routledge is an imprint of the Taylor & Francis Group, an informa business

Cover photo © 1991 Mark A. Anderson

Library of Congress Cataloging-in-Publication Data
Denoon, David.
    Ballistic missile defense in the post–Cold War era / David B.H.
Denoon.
        p.   cm.
    Includes bibliographical references.

    1. Ballistic missile defenses—United States.   2. Antimissile
missiles—United States.   3. World politics—1989–   I. Title.
UG743.D46   1995
358.1'74'0973—dc20   95-12440
                                                          CIP

ISBN 13: 978-0-3670-1702-6 (hbk)
ISBN 13: 978-0-3671-6689-2 (pbk)

Contents

List of Tables and Figures   vii
Preface   ix

1   **The Context**   1

2   **The Changing Meaning of Deterrence**   25

3   **Theater Missile Defense**   53

4   **The Legacy of SDI**   89

5   **National Missile Defense**   119

6   **Missile Defense and U.S. Security Policy**   147

Appendix A: List of the 13 Major Agreements Between the
US and USSR from 1972 to 1979   183
Appendix B: Soviet Approaches to BMD: A Chronology   187
Appendix C: Chemical and Biological Weapons   209
Appendix D: Section of President Reagan's March 23, 1983
Speech Concerning Strategic Defense   211
Bibliography   215
About the Book and Author   230

Tables and Figures

Tables

1.1	Types of BMD	17
3.1	Planned Ballistic Missile Production Ranges	58
3.2	Space Launch Vehicle Producers	60
3.3	Comparing High Explosive Warheads	61
3.4	Potential TMD Systems	83
3.5	TMD/NMD Program Options	86
4.1	Evolution of Space-Based Components	112
5.1	BMD for Defending Population vs. Enhancing Deterrence	122
5.2	Provisions of the Missile Defense Act of 1991	136
5.3	Comparison of Warhead Levels Before and After START I & II	140
5.4	Key Limited Systems Under START I & II	141
6.1	BMD as a Cost-Benefit Problem	149
6.2	Options for BMD Decisions	150
6.3	Present and Potential Future TMD Systems	161
6.4	Policy Changes to Supplement BMD	178

Figures

3.1	TBM Range and Timing Characteristics	55
3.2	North Korea's Nuclear Program	67
3.3	Patterns of Falling Debris from Patriot Interceptors	77
3.4	Active Defense Geometry	80
3.5	Overlap Between Theater and Strategic Systems	81

3.6	Possible Mid-Term TMD System Architecture	84
4.1	Fletcher Report: Phases of Ballistic Missile Trajectories	103
4.2	Fletcher Report: Preliminary Concept for Ballistic Missile Defense During the Boost Phase	104
4.3	Example of Non-Nuclear Ground and Space-Based Architecture	106
4.4	Space Defense System Acquisition Cost	109
4.5	Phase I Architecture with Brilliant Pebbles	116
5.1	Preferred Presidential Policies: Offense vs. Defense	121
5.2	Changes in System Acquisition Cost	127

Preface

Most public commentary on major international issues lags somewhat behind changing realities. At present, the unexpected collapse of the Soviet Union and the end of the Cold War have created a striking change in the world balance of power. Prodigious amounts of effort are being devoted to explaining the resulting shifts in the structure of the international system. This attention to the macro changes in the international environment has led to a neglect of other, less visible changes which could well determine the next important set of adjustments in the global competition for power and influence.

One of the most critical of these new developments has been the growth of the quantity and sophistication of ballistic missile forces in the third world. Ironically, just as many of our foreign policy commentators were concluding that a comprehensive National Missile Defense (NMD) was not necessary for the continental U.S., the missile threat to U.S. forces overseas was growing and the number of third world nations with the capability to develop long-range missiles was also growing. The U.S. Congress did pass the Missile Defense Act of 1991 which actually authorized the development of a limited American NMD system. Unfortunately, Congress never appropriated funds for the procurement of that limited system for the U.S. homeland.

The Clinton Administration has favored new funds for Theater Missile Defense (TMD) that could be used to protect U.S. forces deployed abroad or to protect allies from short- and medium-range missiles; however, it has opposed deployment of any missile defense for the continental U.S. and has turned the strategic defense effort into a purely research program.

The purpose of this volume is three-fold: (1) to explain why ballistic missiles are proliferating and the threat they pose, (2) to identify the principal issues warranting attention in the development of Theater and National Missile Defense, and (3) to demonstrate how adjustments in U.S. strategic thinking and force structure should be made as we plan for the 21st Century.

One of the central themes of this volume is that NMD should be viewed essentially like catastrophic health insurance. All individuals hope to avoid devastating health costs, but many are either unwilling or unable to pay for it. Because of the limited area they protect, TMDs are less controversial, but deciding on NMD is inherently a national issue of major import. Hence, missile defense should be viewed as a "public good" and subject to rational evaluation of its pros and cons. Regrettably, much of the debate about missile defense in the U.S. since 1968 has been highly emotional and infused with various ideological agendas. The attempt here is to present these issues in as direct a manner as possible so that the public and the international community can consider how much it wants to invest in this space-age type of "survival insurance."

*   *   *   *

In working on this manuscript, a large number of individuals and institutions have been helpful to me. Funds from the Sloan Foundation and the American Enterprise Institute first helped launch this effort, and then leaves from New York University and a stay at Stanford University were most helpful in pursuing the project. Particular thanks go to Claude Barfield for early encouragement; and to Frank Miller, Fred Hoffman, Angelo Codevilla, Richard Perle, Ted Postol, and Paul Nitze for insightful comments as this effort got underway. Thanks are also in order for James Abrahamson, Henry Cooper, Keith Payne, Stephen Hadley, Douglas Graham, J.D. Crouch, Steven Cambone, Frank Jenkins and David Wright, who gave me mid-course corrections as the work proceeded. Important data and figures were provided by Thomas Johnson of the Strategic Defense Initiative Organization, and Donald Baucom did a superb historical critique of the developments I was attempting to explain. Oleg Bartrachenko and Li Wei helped research and draft the extensive appendix on Soviet missile defense programs; and B.B. Bell, Harold Brown, Dov Zakheim, Sidney Graybeal and Steven Brams gave me extremely helpful critiques of draft chapters. Special thanks go to Nasrin Abdolali, who not only reviewed the chapters but made numerous helpful comments on the presentation of the data. Finally, Mick Gusinde-Duffy, Eric Wright, and Jim Grode of Westview Press were extremely gracious and supportive as the work proceeded.

David B.H. Denoon
January 1995

1

The Context

For the general public, the Ballistic Missile Defense Era began on January 18, 1991 when a Patriot missile intercepted an incoming Iraqi Scud-B missile over Saudi Arabia. Not only was the intercept at night and a dazzling display of technical virtuosity, but it was recorded on video and replayed numerous times before hundreds of millions of viewers worldwide. The public then recognized that there could be at least limited defenses against ballistic missiles.

Approximately equal numbers of Scuds were fired at Saudi Arabia and Israel.[1] In Saudi Arabia, one of the Scuds hit an American barracks in Daharan, killing 36 soldiers, while in Israel there was widespread light damage from falling debris, especially during the initial week of attacks between January 18-25, 1991.[2]

Nevertheless, there is broad agreement that the availability of some ballistic missile defense (BMD), such as the Patriots, calmed the public in both Saudi Arabia and Israel.[3] Moreover, there were no substantial civilian casualties in either Israel or Saudi Arabia. This is in sharp contrast to the

---

[1] According to final Department of Defense figures, Iraq fired 42 Scuds at Israel and 46 at Saudi Arabia and Bahrain. See, *Conduct of the Persian Gulf War, Final Report to Congress*, (Washington D.C.: Department of Defense, April 1992), p. 168.

[2] Postol, T., "Lessons of the Gulf War Experience with Patriot," *International Security*, Winter 1991-92, Vol. 16, No. 3, p. 141.

[3] The intense controversy over the actual effectiveness of the Patriots in the Gulf War will be discussed in a separate section of Chapter 3.

Iran-Iraq War where Baghdad's Scuds caused over 1100 dead and 4000 wounded in Tehran.[4] The Patriots also played a key role in the political context of the Gulf War, as they were an important factor in the Israeli leadership's decision not to retaliate and enter the war. This, in turn, made it much more likely that the Arab nations would continue to participate in the Allied Coalition against Iraq.

The Patriot missile being used in 1991 was the PAC-2, an upgraded version of a missile that was originally designed for use against aircraft. It was meant for small area defense of military positions. The Patriot was never intended for defense of cities or for dealing with Iraqi-modified Scuds which had such bad metallurgy that they were literally disintegrating into smaller pieces as they descended toward their targets.

It is important to make a distinction between different types of BMDs.[5] The Patriots in use during the Persian Gulf War were for dealing with short-range, theater missiles, and were, therefore, Theater Missile Defenses (TMDs). Most of the debate regarding the Anti-Ballistic Missile Treaty of 1972 and the Strategic Defense Initiative in the 1980s concerned interceptors for missiles that could travel 5000 KM or more. Hence defenses against longer range missiles have been called strategic defenses or National Missile Defenses (NMD). U.S. TMDs have only been operational in the last decade; and, since the technical problems of TMD and NMD are quite different, caution in necessary in making generalizations about the two types of BMDs. Nevertheless, the general public frequently does not follow these differences, and the support for or opposition to BMD programs often comes from failing to distinguish between the specific challenges facing TMD and NMD efforts.

Twenty five years earlier, in February 1966, when Secretary of Defense McNamara authorized the first research and development funds for the SAM-D missile (which ultimately became the Patriot), he had no idea that he would

---

[4]   Carus, W.S. and J. Bermudez, "Iraq's Al-Hussein Missile Programme," *Jane's Soviet Intelligence Review*, June 1990.

[5]   In this volume, there were two choices of terminology which were made to clarify the discussion. First, as will be explained in detail in Chapter 2, the word "Deterrence," with a capital D, is a generic term to describe the overall military strategy of the U.S. from 1962 to the present. Thus, "Deterrence" is broader in scope than "mutual assured destruction." Second, the term National Missile Defense (NMD) has been substituted for "strategic defense." Although NMD was not used as a term until the 1990s and it is somewhat artificial to refer to "NMD" when reviewing the 1980s debates, this was done to maintain consistency in terminology throughout the book.

be laying the groundwork for a system which would undercut his tenacious and oft-repeated preference for deterrence over BMD.[6] In fact, McNamara and much of the academic community became bitter critics of NMD when the Reagan Administration was trying to develop support for the Strategic Defense Initiative after 1983.[7]

Most of the people who saw the success of the Patriots[8] in 1991 had no idea that defense specialists and various activists had been debating the feasibility and desirability of BMD since 1958.[9] Also, the American public would probably be aghast if it fully understood that three decades of political leaders including President Clinton, had agreed to a defense strategy that had no active BMD for the continental U.S.[10]

U.S. strategic policy has been and still is based on the presumption that NMD or "Defense" is less desirable than "Deterrence."[11] The essential principle of "Deterrence" is that all opponents should recognize, if they attack, they will be hit with an unacceptably large counter-strike. The key element in ensuring that your opponent knows you have "an assured second strike" is reliable military systems that can survive an attack and still deliver a devastating response.

The effectiveness of the Patriots in the Gulf War led to a surge in public support for all missile defenses during 1991. This, along with other factors

---

[6] For an example of McNamara's 1980s argument for "Deterrence," see, Bundy, McG., G. Kennan, R. McNamara, and G. Smith, "The President's Choice: Star Wars or Arms Control," *Foreign Affairs*, Winter 1984/85, Vol. 63, No. 2, pp. 264-278.

[7] For several examples, see: H. Bethe, R. Garwin,, K. Gottfried, and H. Kendall, "Space Based Ballistic Missile Defense," *Scientific American*, Vol. 251, No. 4, October 1984, pp. 39-44, and W. Panofsky, "The Strategic Defense Initiative: Perception vs. Reality," *Physics Today*, June 1985, pp. 33-45.

[8] As stated, the Patriots used in 1991 were PAC-2s which stands for Patriot Anti-Tactical Missile Capability #2. The final development of the PAC-2 took place during the Summer and Fall of 1990 as new software and guidance technology were added to the previous version, PAC-1.

[9] The NIKE II study by Bell Labs in 1955 did simulations of missile defense effectiveness, but it was classified and the public debate did not surface until the Soviet Sputnik satellite was launched in October 1957.

[10] There was a very brief period of 1974-76 when the U.S. did have an operational BMD at Grand Forks, North Dakota; but it was designed to protect missile fields and was not capable of defending any substantial part of the population.

[11] "Deterrence" will be used here with a capital "D" to refer to the form of deterrence that the U.S. chose during the past 30 years, i.e. no deployed strategic defense capability at all.

like uncertainty over events in the former Soviet Union, led the Congress to pass the Missile Defense Act of 1991. The Congressional action authorized the development of further Theater Missile Defense (TMD) systems like Patriot and a National Missile Defense (NMD) system that could intercept and destroy a limited attack, using up to 100 ground-based interceptors.

Ironically, what many in the Congress and press did not know was that the technical problems of TMD are significantly easier to overcome than the problems of national defense. This is because short and medium range missiles spend much more of their flight time in the atmosphere and leave a thermal image that can be followed through most of their trajectory. Intercontinental ballistic missiles (ICBMs), on the other hand, spend most of their time in the cold, black parts of space outside the atmosphere which makes them hard to track. Additionally, most ICBMs reenter the atmosphere going twice or three times as fast as theater missiles.

Thus, we have a situation where the public was impressed with one type of BMD (Patriot) and this swayed some skeptics in Congress.[12] Yet, there is little general understanding of the myriad engineering and production problems that remain to be solved in developing an effective NMD. Nor does most of the public know how bitter the feuds have been between those favoring offensive systems ("Deterrence") versus those advocating various BMDs ("Defense").

The intensity of the opposition to BMD has been extraordinary. The academic and media campaign against the Nixon Administration's proposal to build the Safeguard system in 1969 was similar in tone and involved many of the same individuals[13] as the onslaught against the Strategic Defense Initiative (SDI) in the mid-1980s.[14] Opponents of NMD have been very worried that the public would draw conclusions about strategic or national defense from the 1991 Patriot performance; and some have even gone so far as to argue that soldiers and citizens of Saudi Arabia and Israel would have been better off if the Patriots had not been used.[15]

---

[12] Broad, W., "Congress Preparing to Quietly Approve an Anti-missile Plan," *New York Times*, November 18, 1991, p. A-1.

[13] See, A. Chayes and J. Weisner, eds., *ABM: An Evaluation of the Decision to Deploy an ABM System*, (New York: Harper and Row, 1969).

[14] See, J. Tirman, ed., *The Fallacy of Star Wars*, (New York: Vantage Books for the Union of Concerned Scientists, 1969).

[15] Postol, T., "Lessons from the Gulf War Patriot Experience: A Technical Perspective," *Mimeo.*, Testimony before the House Armed Services Committee, April 16, 1991.

Overview of the Current Debate

Why has this issue generated such intensity and invective over such a long period of time? There are clearly a host of factors and they have varied in the three decades of controversy. Nevertheless, there are some fundamental patterns and splits among the protagonists in the debate. There have been and are deep differences over: (1) ideology, (2) attitudes toward defense spending and how best to deal with the Soviet / Russian missile force, (3) bureaucratic gains or losses for the Army, Navy, or Air Force, (4) the implications for arms control, and (5) differences about the underlying strategic assumptions inherent in "Deterrence" rather than "Defense."

On balance, however, it is fair to say that supporters of BMD have usually been ideologically conservative, often willing to spend more on defense, skeptical of Soviet / Russian claims and actions, less concerned about arms control, and unconvinced that relying entirely upon offense is wise. Conversely, supporters of "Deterrence" and opponents of BMD have frequently been politically liberal, less skeptical of Moscow's behavior, critics of defense spending in general, and often avid devotees of arms control.

Hence, during the Cold War, and especially in the mid-1960s to the mid-1980s when the USSR was steadily expanding its defense spending, many critics argued that a U.S. NMD would simply lead to a heightened arms competition among the super-powers. In that period, supporters of NMD faced a major hurdle. Along with demonstrating that the NMD system could actually intercept and destroy hundreds or possibly thousands of incoming re-entry vehicles, they were asked to show that the USSR would not respond in some counter-productive fashion. A large debate then developed over estimating what counter-measures the USSR might take if the U.S. built an NMD. Thus, advocates of NMD had to show both that missile defense would be *technically feasible* and that there would be *benign reactions to deployment*. This standard was enough to thwart the construction of any large-scale NMD.[16]

President Reagan's speech of March 23, 1983, advocating the development of a strategic defense system (see Appendix D), was intended to launch a revolution in U.S. national security policy. Although there is controversy about whether the White House relied on objective scientific advice,[17] there is

---

[16] The U.S. opposition to BMD produced a curious situation where the U.S. gave up its only operational BMD, whereas the USSR / Russia has maintained a functioning BMD system around Moscow for three decades.

[17] Lakoff, S. and H. York, *A Shield in Space? Technology, Politics and the Strategic Defense Initiative*, (Berkeley, Cal.: University of California Press, 1989), pp. 9-15.

little doubt that the President considered "Defense" both politically and morally preferable to "Deterrence."

Nonetheless, despite important technical advances in the period between 1969 and 1983,[18] President Reagan was initially still dealing with an antagonistic Soviet Union led by General Secretary Yuri Andropov, a former head of the KGB. So, in addition to all the scientific debate that the Strategic Defense Initiative (SDI) faced, there was still the dilemma of how the USSR would respond. Andropov played this to the hilt, stating what a grave threat SDI was and how destabilizing the situation would be as the U.S. was preparing to develop a "first-strike capability." This sent much of the American arms control community and the media into high states of agitation.

One of the ironies of the supposed stark difference between "Deterrence" and "Defense" is that the U.S. has for decades spent substantial sums on "passive defense" without causing trauma in the USSR or among opponents of BMD. For example, the U.S. reinforces missile silos with concrete and various shock absorption devices so they can survive near-miss nuclear explosions. Also, vast expenditures have been made to make submarines hard to detect and to keep certain aircraft on alert so they can take off on warning of a missile attack. These passive measures are all intended to reinforce "Deterrence." Some have argued that, without these passive defense measures, the U.S. strategic retaliatory forces would be "more of an invitation to attack than a deterrent against it."[19]

BMD opponents have been reluctant for a long time to acknowledge the similarity between active and passive defenses, yet still treat passive defenses as acceptable and active defense as anathema. However, there have been four additional changes on the world scene since 1983 that fundamentally alter the previous calculus for the U.S. about the desirability of BMD. These critical changes are:

1) The unanticipated and rapid decline of Soviet power,
2) The signing of the START I and II agreements,

---

[18] The three principal technical advances which made *some* scientists optimistic about strategic defense effectiveness were improvements in: (1) computing speed, (2) signal processing, and (3) miniaturization. These advances made missile interceptors more accurate, and made small, space-based weapons possible for use during the "boost-phase" of a missile attack. These issues will be discussed in detail in Chapter 3.

[19] Perle, R., "Statement Before the Committee on Armed Services," U.S. House of Representatives, April 16, 1991.

3) The spread of chemical, biological and nuclear weapons technology, and

4) The rise of a third world ballistic missile threat and leaders who were not convinced that U.S. would use its nuclear weapons.

In 1986, when the Union of Concerned Scientists drew up their list of countermeasures that the Soviet Union might take if the U.S. deployed strategic defenses, it was a long and imposing set of potential responses.[20] It was argued that the USSR might totally replace its missile force, employing new methods that would thwart space-based lasers.

The Union of Concerned Scientists claimed that the Soviets might do any or all of the following: (1) deploy fast-burn boosters that would not be firing outside the atmosphere, thus making thermal imaging harder, (2) actually spin missiles, so that the lasers would not be able to penetrate the missile skin quickly, (3) harden the exterior skin of missiles to resist lasers, (4) develop decoys that had physical characteristics like actual warheads, and (5) change the trajectories of missiles so that they would circle the earth or follow indirect routes in ways unanticipated by the defense. Although there were legitimate doubts at the time whether the Soviet Union really had the technical capacity to make these counter-measures work, each of the steps was physically possible and could not be dismissed entirely if the Soviet Union was really willing to completely replace its missile force.

Yet, today we can reasonably assume that the type of counter-measures that the Union of Concerned Scientists envisioned would be so expensive that no country is currently prepared to undertake them. Certainly redesigning and rebuilding its entire missile force would be an extremely controversial step for Russia to take and not one that a President with low popularity ratings like Yeltsin is likely to try. Moreover, all the residual states of the former Soviet Union are in perilous economic condition and they would risk having external aid cut off if they initiated a new arms build-up. Although China could potentially build a new, sophisticated missile force, there are no current indications that Beijing has such plans. Hence, the type of "action-reaction" sequence from major potential opponents, which the arms control community postulated as a key danger from NMD deployments, is a low probability outcome now.[21]

---

[20] For a full discussion of the Union of Concerned Scientists' views on counter-measures, see R. Garwin, "The Soviet Response," Chapt. 6 in J. Tirman, ed., *Empty Promise - The Growing Case Against Star Wars*, (Boston: Beacon Press, 1986), pp. 129-146.

[21] Some opponents of improved TMDs argue now that even TMDs could intercept a sufficient number of ICBMs to encourage China to upgrade its offensive mis-

Also, since the late 1980s, there have been a series of accommodating and mutually cooperative moves that the former Soviet Union and the U.S. have made which make it plausible for them to work together on considering alternatives to "Deterrence." In 1990 President Gorbachev took some unilateral moves that significantly reduced the military threat to the U.S. and its NATO allies. He announced military manpower cuts beyond those specified in the Mutual and Balance Force Reductions (MBFR) agreement, oversaw the dissolution of the Warsaw Pact, and agreed to German unification with the former East Germany as part of NATO. That he made these moves out of economic and political necessity does not diminish the military implications of the action. The U.S. and Western Europe are demonstrably safer places today than they were in 1983 because of *unilateral* actions by Gorbachev.

President Bush reciprocated with some important American moves. On September 27, 1991, he announced that the U.S. would: take part of the U.S. strategic bomber force off alert status, begin disassembling and eliminating all short-range nuclear weapons, quit deploying all air-launched and sea-launched nuclear cruise missiles, and cancel the development of the mobile M-X and Midgetman missiles.

It was disclosed two weeks later that the U.S. had proposed to Soviet negotiators in Geneva that strict limits be put on the extent of strategic defense systems if the USSR would agree to modify the ABM Treaty to permit deployment of: (i) a specified number of ground-based interceptors, and (ii) a fixed number of space-based systems.[22] Although the Soviets did not agree to this Bush proposal and shortly thereafter the Soviet Union split into its current constituent states, President Yeltsin committed Russia to the second Strategic Arms Reduction Treaty (START II) and there have even been proposals from Moscow that Russia join NATO.[23]

Thus, some of these U.S. and Russian moves were in a formal, negotiated arms control setting and others were not. We may be moving away from a situation where there is a compulsive concern with "mirror imaging" and the two sides may pursue separate and, in certain cases, unrelated defense policies.

This does not mean, however, that there is no danger to the United States

---

(Continued)

sile capabilities. See Chapter 6.

[22] Rosenthal, A., "U.S. Offers to Negotiate on 'Star Wars'," *New York Times*, October 16, 1991, p. A-3.

[23] Friedman, T. "Cold War Without End," *New York Times Magazine*, August 22, 1993, p. 30.

from ballistic missiles. An accidental launch is very improbable but not statistically impossible; and there is some chance that a submarine commander might circumvent the controls and fire one or more missiles. This risk of accidental or limited attacks is one of the reasons that the Bush Administration changed the goals of the Strategic Defense Initiative and urged the development of a system it called Global Protection Against Limited Strikes (GPALS).[24]

The trends listed above are mostly positive and certainly ones that will give both Russia and the U.S. more flexibility in their defense policy choices. BMD in the middle 1990s has both positive and negative features, but it would be outmoded and unnecessarily rigid to assume that choices about BMD today are constrained in the same way as they were when President Reagan first proposed the SDI. The end of the Cold War has created new complexities, but it has eased rather than increased constraints on BMD policy.

The most important emerging factor which has created support for BMD is the growth of ballistic missile threats from the third world.[25] Although China is the only developing country with ICBMs capable of hitting the U.S. today, the Persian Gulf War made it obvious that allies of the U.S. as well as American troops deployed overseas are *already* threatened by ballistic missiles. Also, over time, the number of countries that have both missiles and nuclear weapons technology will grow.[26] Most Americans would favor protecting U.S. forces deployed overseas with TMDs, but it is a more complex, technological and diplomatic issue to decide if NMD is necessary now.

The question then to ask is: how much longer should the U.S. stick with "Deterrence" as the underlying concept for its strategic forces? This may have been a reasonable way to deal with the U.S.-Soviet rivalry, but threatening nuclear retaliation, clearly, will deal with only a limited number of circumstances. During 1989, as Soviet power was crumbling, there were a num-

---

[24] For a summary of the changes between the Reagan Administration goals for SDI and the Bush Administration goals for GPALS, see K. Payne, *Missile Defense in the 21st Century: Protection Against Limited Threats*, (Boulder, Colo.: Westview Press, 1991).

[25] For detailed analyses of third world missile programs, see W. Seth Carus, *Ballistic Missiles in Modern Conflict*, (New York: Praeger, 1991) and J. Nolan, *Trappings of Power - Ballistic Missiles in the Third World*, (Washington, D.C.: Brookings Institution, 1991).

[26] Spector, L., *Nuclear Ambitions - The Spread of Nuclear Weapons 1989-1990*, (Boulder, Colo.: Westview Press, 1990).

ber of fanciful suggestions that, as the Cold War faded, conflicts around the globe would also decline and there might even be an "end to history" as people ran out of compelling ideas to argue about.[27] The Gulf War, the continued fighting in Somalia and Angola, and the Bosnian conflict all demonstrate that wars have not faded from the scene.

Nevertheless, if managing super-power rivalry is a smaller part of U.S. defense problems and dealing with regional and ethnic conflicts is a growing challenge, then the paradigm on which "Deterrence" is based will lose adherents. Much as Newtownian physics and pure Keynesian economics lost appeal, pure "Deterrence" will lose its allure.[28] Long-term advocates of assured destruction will not easily switch their views, and decisions are likely to be incremental. For example, President Bush's canceling of the mobile M-X and Midgetman, and deep Congressional skepticism about the need for a large number of B-2 Stealth bombers, appear to be simply pragmatic adjustments to a changed threat. Yet, this may also mean that fewer traditional offensive systems will be available for actual procurement. In the future, the U.S. might then shift to more non-nuclear ordnance with precision-guided munitions (PGMs) using terminally-guided warheads (TGWs).[29] Without question this will change strategic options.

As confidence in "Deterrence" as an organizing concept fades, some will be searching for a new orthodoxy. However, BMD is only a cluster of technologies. It is not a theory or an all-encompassing view of how to design policy. Doubtless there will be those advocating a purely defensive strategy for the U.S., while others may want a heavy BMD while continuing to maintain an extensive offensive capability.

There are obviously a range of strategic options incorporating various levels of BMD that could be pursued. This is not a book on strategy in general, and it will not attempt to answer precisely how much BMD the U.S. should develop. However, it is a central theme of this volume that BMD will play not only a role, but an increasingly important role in U.S. defense strategy in the coming decades.

---

[27] Fukuyama, F., "The End of History," *The National Interest*, Summer 1989, No. 16, pp. 3-18.

[28] For the best overall description of how a paradigm gains followers, develops into a school, and then declines, see T. Kuhn, *The Structure of Scientific Revolutions*. 2nd. ed., (Chicago, Ill: University of Chicago Press, 1970).

[29] Nitze, P., "After Iraq, Nukes Can Be Junked," *Wall Street Journal*, December 24, 1991, p. 15.

Changing Views on BMD

Given the strong Congressional resistance to BMD in the 1969-1990 period, what led to the reversal of sentiment and the approval of BMD in concept with the Senate and House approval of the Missile Defense Act of 1991?

Several factors appear to have contributed to the passage of the legislation. The coup attempt against President Gorbachev in August 1991 brought into full view the potential instability in the former Soviet Union and the danger that might come if a radical group were to get control of nuclear armed missiles. Second, the success of the Patriots in the Gulf War showed that, at least TMD could work, and that real dangers would result if BMD was ignored. Third, several delegations of U.S. Congressmen and Senators had been to Moscow in 1991 and found that the Russians themselves were worried about third world ballistic missiles and were no longer as rigid as before about rejecting amendments to the Anti-Ballistic Missile (ABM) Treaty.

As noted above, Congress then approved an interesting compromise: (1) active research would continue on a broad range of BMD technologies, (2) the Department of Defense could actually begin the procurement process for a treaty compliant NMD system that could deploy up to 100 interceptors, and (3) rapid development of upgraded TMD that would have capabilities substantially greater than the Patriot PAC-2s used in the Persian Gulf War.

The political support for the Missile Defense Act of 1991 was a diverse coalition. In the Senate, two moderate Republicans, Lugar and Cohen, joined with a conservative, Warner, to urge support for BMD. Those three Republicans had made a joint trip to Moscow and were convinced that the Russians were ready to rethink their position on expanded BMD. On the Democratic side, Senator Nunn did not want this to become entirely a Republican initiative; and, despite his mixed support for BMD at different times, stressed that it was important for the country to have a long-term plan for how TMDs and NMD should be developed. This earned Nunn considerable criticism from opponents of BMD.[30]

In addition, even the more liberal Democrats in the House were worried about control of Soviet missiles during the break-up of the Soviet Union. Also, though the overall military campaign against Saddam Hussein had been a lopsided victory, the U.S. Air Force had not been notably successful at finding and destroying the Iraqi mobile Scuds. Thus, the Congressional coalition

---

[30] See, the *New York Times* editorial, "Mr. Nunn's Rash Rush to ABMs," July 29, 1991.

which ultimately supported the Missile Defense Act included some who were primarily worried about TMD and others who focused on NMD of the U.S. homeland. The combination of their concerns produced the first authorization for procurement of a BMD system since 1969.

The period between 1991 and 1993 produced mixed messages on BMD. The management and priorities of the Strategic Defense Initiative Office (SDIO) were criticized. For example, Senator Nunn expressed dissatisfaction with the pace at which the Department of Defense moved on TMD,[31] and the Congress was never willing to endorse the small, space-based interceptor rockets (named Brilliant Pebbles) which the Bush Administration favored for NMD of U.S. territory.[32] Evidence of the growing concern about proliferation of nuclear weapons and ballistic missiles in the third world is broader today than in 1991.

The U.S. has made special efforts to convince North Korea to adhere to the 1963 Nuclear Non-Proliferation Treaty,[33] stop Russia from exporting space launch technology to India, and put pressure on China to stop selling M-11 missile parts to Pakistan.[34] The Clinton Administration apparently favors a broad arms control plan which would restrict sales of highly enriched uranium and plutonium.[35] The Clinton Administration has also taken a very explicit position on the ABM Treaty, advocating the "narrow interpretation" of those programs which conform to the Treaty. The Arms Control and Disarmament Agency has stated in a letter to Congress that: "the ABM Treaty prohibits the development, testing, and deployment of sea-based, air-based, space-based, and mobile land-based ABM systems."[36] Additionally, in a symbolic step, former Secretary of Defense Aspin changed the name of the Pentagon office

---

[31] See, H. Cooper, "End of Tour Report," *Mimeo.*, (Washington, D.C.: U.S. Department of Defense, January 20, 1993).

[32] The Bush Administration also favored having ground-based interceptors, but it was the push for Brilliant Pebbles that aroused Congressional resistance.

[33] Jehl, D., "U.S. Outlines Concern Over North Korean A-Arms," *New York Times*, February 25, 1993, p. A-7.

[34] Holmes, S., "China Denies Violating Pact by Selling Arms to Pakistan," *New York Times*, July 26, 1993, p. A-2.

[35] Gordon, M., "U.S. Hopes to Curb A-Arms By Restricting Fuel Output," *New York Times*, July 28, 1993.

[36] The letter was from Thomas Graham, the Acting Director of the U.S. Arms Control and Disarmament Agency, to Senator Claiborne Pell, Chairman of the Senate Foreign Relations Committee. See, T. Friedman, "U.S. Formally Rejects 'Star Wars' in ABM Treaty," *New York Times*, July 15, 1993, p. A-6.

dealing with missile defense from the Strategic Defense Initiative Office to the Ballistic Missile Defense Office (BMDO).

A large controversy regarding BMD has erupted recently, however, and it regards a 1984 test done by the SDIO when it was directed by General James Abrahamson. In 1984 the SDIO announced that, in a test on the Kwajalen Range in the Pacific, an interceptor rocket had successfully hit and destroyed an incoming ICBM fired from the West Coast of the U.S. This was called the "Homing Overlay Experiment" and was seen as a major success for non-nuclear, ground-based interceptors.

Some analysts have claimed that the chances of hitting the incoming ICBM were greatly increased by putting a beacon on it, so that the interceptor would have accurate coordinates to aim for.[37] After reviewing the history of the experiment, former Secretary Aspin said that the *New York Times* story was false. Aspin directly stated that the beacon had nothing to do with guiding the intercepting rocket to the incoming ICBM.

Nevertheless, some critics have claimed that the Department of Defense (DOD) presentations were designed to give an over-optimistic assessment of U.S. progress on strategic defense. This may have contributed to the USSR spending unnecessary money on its military programs. Obscuring the actual details of the experiment also appears to have misled some members of Congress.

Regardless of the final judgment about the actual performance in the Homing Overlay Experiment, skepticism about the integrity of the SDIO procedures had a negative effect on government credibility. Not only will this add to cynicism about the validity of U.S. BMD tests, but there will be caution regarding claims by future managers of the programs.

In September 1993, former Secretary of Defense Les Aspin completed a broad study of U.S. strategy and programs, calling it the "Bottom-Up Review."[38] In the conventional forces area, the Aspin recommendations were remarkably close to those made by the Bush Administration in its "Base Force" plan.[39] For example, Aspin suggested reducing the Army to 10 active

---

[37] Weiner, T., "Lies and Rigged 'Star Wars' Test Fooled the Kremlin and Congress," *New York Times*, August 18, 1993, p. A-1. See Footnote 53 in Chapter 4 for further debate on this issue in 1994.

[38] Aspin, L., "The Bottom-Up Review: Forces For A New Era," *Mimeo.*, (Washington, D.C.: U.S. Department of Defense, September 1, 1993).

[39] Gordon, M., "Military Plan Would Cut Forces But Have Them Ready for 2 Wars - Clinton Strategy Endorses Bush's Basic Doctrine," *New York Times*, September 2, 1993, p. A-1.

duty divisions instead of 12 but urged more Marines than Bush recommended. Similarly, the active duty Navy and Air Force are to be slightly smaller than Bush advocated, but only one carrier battle group less in the Navy and only one fighter wing less in the Air Force.

On the other hand, Aspin recommended major changes in the direction and scope of the Defense Department's BMD programs. For the five year period, Fiscal Year 1995-99, Aspin favored cutting spending from the $39 billion level recommended by Bush to $18 billion. In addition, the Bottom-Up Review urged a fundamental redirection of efforts within the BMD area.

The Clinton Administration focus will be on TMD and the development of a core program, including: upgrading of Patriot, proceeding with the development of the Theater High Altitude Area Defense (THAAD) missile, upgrading naval TMDs and improving battle management and communications systems. Only 1/6 of the funding will go for national or strategic defense. All of these choices will be analyzed in greater detail in the following chapters, but there is little doubt that the Clinton Administration prefers smaller BMD programs and that the principal goals will be to improve theater rather than strategic systems.[40] The newly elected Republican Congress may attempt to change these priorities.

Goals of the Book

As stated, this is not a book about strategy per se but about how certain military technologies can be used and their implications for U.S. defense policy. NMD has been thwarted for three decades for a variety of reasons; but, now TMDs are close to being deployed as an operational part of U.S. forces, there is a need to see its history in context and attempt to define its significance.

There are three principal goals of this volume:

1) To explain why short and medium range missiles are proliferating (just at a time when the START talks have led to commitments for significant reductions of U.S. and Russian ICBMs), and to identify the range of problems that this poses for U.S. defense policy.

2) To analyze how the U.S. focus on offensive systems was reinforced

---

[40] For a third view, which concisely links recommended U.S. deployments to a political military strategy, yet is distinct from both the Bush and Clinton Administration approaches, see Robert Art, "A U.S. Military Strategy for the 1990s: Reassurance Without Dominance," *Survival*, Winter 1992-93, Vol. 43, No. 4, pp. 3-23.

by: (a) the long-standing Air Force and Navy commitment to the Triad, and (b) the academic community's commitment to "Deterrence" and arms control.

3) To demonstrate how both U.S. doctrine and force structure need to be modified to deal with the changing national security environment, and that these adjustments are different from just a down-sizing of the current forces.

Definitions and Main Assumptions

The essence of "Deterrence" has been defined above as keeping your opponent from attacking by assuring that you have the ability to survive the attack and deliver a devastating response. This concept was developed in the late 1950s by a combination of defense specialists, game theorists,[41] and economists who were trying to optimize military expenditures by selecting the lowest cost force structure which would still ensure national security.[42] "Deterrence" became official U.S. Government policy under Secretary of Defense McNamara in 1962,[43] and updated variants of the policy are still in effect.

Because most scientists in the late 1950s were convinced that BMD systems of that era could be overwhelmed, defense was not the critical variable. The issue became: how much offense do you need? The answer, according to this view was: you need enough so that your opponent will never attack you in the first place. The designers of "Deterrence" then concluded that you needed only enough to inflict an unacceptable level of damage on your opponent if he were so unwise as to attack.

Therefore, instead of concentrating on defenses to limit the damage of an opponent's attack, advocates of "Deterrence" urged having only a survivable "second-strike capability" should the U.S. be hit. So, national security was to be ensured by convincing all opponents that an attack on the U.S. might inflict

---

[41] For a clear presentation of the early game theory concepts in "Deterrence," see T. Schelling, *The Strategy of Conflict*, (New York: Oxford University Press, 1960).

[42] For a history of the organizations and key individuals involved in this effort, see, F. Kaplan, *Wizards of Armageddon*, (New York: Simon & Schuster, 1983).

[43] An unclassified version of the policy was presented by Secretary McNamara at his University of Michigan commencement address in June 1962.

damage, but that there was enough diversity and range in American forces to retaliate and impose intolerable levels of damage on the attacker. Unquestionably, there could be debate about (a) how large the second strike should be or (b) how to disperse, disguise, or protect the retaliatory force, yet the goal was to convince any and all antagonists not to attack at all.

There was a principal corollary to the "Deterrence" view: avoiding excessive capabilities of your own so that you did not unnecessarily threaten others. "Deterrence" thus became inextricably linked with arms control.[44] Some defenders of "Deterrence" as a concept treat it almost with reverence and see preserving the ABM Treaty and the related arms control agreements as a goal that should sustain supporters even into the 21st Century.[45] It is curious that many defenders of "Deterrence" see the end of the Cold War as requiring dramatic changes in the rest of U.S. forces but prefer to maintain assured destruction as the strategic concept of choice.

From the beginning, however, there were two essential questions that "Deterrence" doctrine could not answer: (1) Why is it politically or morally preferable to threaten to annihilate your opponent rather than defending your own citizens?[46] and (2) If you have only offensive weapons, what do you do for your own people if "Deterrence" fails?

Over the past three decades, there has been a sizable intellectual exchange about the limitations of "Deterrence." Although advocates of BMD have on two occasions convinced the President to shift at least to some "Defense," this has never gotten full-scale approval from Congress. At present, supporters of "Defense" form a broad spectrum. They vary from those advocating a comprehensive population defense[47] to others who see BMD fulfilling a range of roles from supplementing "Deterrence" to "point defense" of military facilities to various population defense options.[48]

---

[44] These links are laid out in Chapt. 3 of C. Blacker and C. Duffy, eds., *International Arms Control - Issues and Agreements*, (Stanford, Cal.: Stanford University Press, 1984).

[45] For an example of this view, see A. Chayes and P. Doty, eds. *Defending Deterrence - Managing the ABM Treaty into the 21st Century*, (Washington, D.C.: Pergamon-Brassey's, 1989).

[46] For an overview of the ethical debates regarding deterrence, see, R. Hardin, J. Mearsheimer, G. Dworkin, and R. Goodin, eds., *Nuclear Deterrence - Ethics and Strategy*, (Chicago, Ill.: University of Chicago Press, 1985).

[47] Worden, S., *SDI and the Alternatives*, (Washington, D.C.: National Defense University Press, 1991).

[48] For an overview of different defense options, see F. Hoffman, A. Wohlstetter, and D. Yost, *Swords and Shields - NATO, the USSR and New Choices for Long-*

Since there are numerous types of BMD and various ways that defensive strategies could be combined with offense, it is harder to define a precise doctrine of "Defense" equivalent to "Deterrence." Nevertheless, there are some central views that most advocates of BMD share: (1) "Deterrence" provides no guidance on how to conduct a war if one breaks out and it is irresponsible to leave either civilians or troops unprotected in these circumstances; (2) Passive defense and defenses against aircraft have been accepted throughout the "Deterrence Era," so why make an artificial distinction about BMD? and (3) Although it may have been a pragmatic decision to accept mutual vulnerability with the USSR during the Cold War, there is no necessary reason to exalt in it or accept it when the circumstances no longer warrant it. Table 1.1 below outlines the three functions currently envisaged for BMD and the U.S. systems that might be used for those roles.

TABLE 1.1 Types of BMD

Function	Type of Protection	U.S. System
Theater Defense	Point or small area (25-50 sq. miles)	Patriot PAC-3, ERINT [a], THAAD [b]
Accidental or Unauthorized Launch	Limited Protection for Continental U.S.	GBI [a]
National Defense	Thin Population defense Long-term future	GBI GPALS with BP Directed Energy [a]

a. GBI is Ground Based Interceptor, ERINT is Extended Range Interceptor, and Directed Energy refers to lasers or particle beams.
b. THAAD's capabilities are classified but it is an upper-tier interceptor, designed to defend a much wider area than Patriot PAC-3 or ERINT.

(Continued)
Range Offense and Defense, (Lexington, Mass.: Lexington Books, 1987).

In Chapters 3 and 4 we will have an extensive discussion of the pros and cons of different BMD systems.[49] What follows are three principal assumptions on which this book is based. The assumptions will not be analyzed, but are meant to provide the background for the propositions and basic argument of the volume.

Assumption 1

Estimating the probability of a ballistic missile attack on the U.S. homeland or American troops overseas is an inherently imprecise exercise. However, all "decisions between national security postures are bets about how best to control or influence rather distant futures."[50] So to make rational decisions about whether to proceed with BMD, it is necessary to make at least crude judgments about the relative likelihood of future events. Obviously, just because missile attacks have already occurred on U.S. forces abroad does not mean that U.S. soil will ever be attacked; but, conversely, the possibility should not be dismissed entirely.

Assumption 2

As the technical competence of the Less Developed Countries (LDCs) improves, it will lead to increasingly sophisticated weaponry being in the hands of a larger number of countries. This will widen the range of the weapons and the number of countries that are potential ballistic missile targets. Since the spread of technology is inevitable, there are at least two plausible courses of action: (a) attempting to slow the sale of actual weapons and the diffusion of technologies critical for weapons production, and (b) preparing to defend against the spreading threat.

Assumption 3

In the post World War II period the U.S. has provided security protection for many allies and non-allied countries but only a few of these countries con-

---

[49] For descriptions of the different BMD systems listed in Table 1 and their performance characteristics, see, "The New Focus for SDI: GPALS," *Mimeo.*, (Washington, D.C.: U.S. Department of Defense, SDIO, June 6, 1991).

[50] Bobrow, D., "Improving the Bases for Decision," Chapter 1 in D. Bobrow, ed., *Weapons System Decisions - Political and Psychological Perspectives on Continental Defense*, (New York: Praeger, 1969), p. 4.

tributed significantly to U.S. national security.[51] As the range and sophistication of ballistic missile threats increase, the U.S. will need a way to limit the extent of its commitments. "Defense" rather than "Deterrence" makes it far more feasible to limit the extent of U.S. engagement because American commanders may wish to defend troops but not necessarily retaliate as "Deterrence" would require.

Basic Arguments and Propositions of the Book

The basic argument of this volume is that the U.S. needs to reconsider some of its most fundamental security policies. "Deterrence" has provided a means for dealing with the U.S.-Soviet rivalry during the Cold War, but it is too broad and unfocused a doctrine to be an adequate guide as the nature of the threats change. The discussion in the chapters below will be analyzing and evaluating the following eight propositions.

1) The dominant strategic concept for the U.S. in the past 30 years has been "Deterrence." Most American policy-makers and defense analysts have favored offensive systems that would provide "Mutual Assured Destruction" as the best guarantee against any opponent launching an initial attack on U.S. soil and American allies.

2) Interest in missile defenses has surged and waned during the past three decades, but a combination of three main factors has inhibited the move toward a ballistic missile defense capability: (a) the momentum of Air Force and Navy procurement and force structure, (b) the technical superiority of offensive systems over defensive systems during the 1950s-1980s, and (c) the strong commitment to "Deterrence" as the most cost-effective way to influence Soviet military and civilian decision-makers.

3) There have been two attempts by recent Presidents to challenge the orthodoxy of "Deterrence." President Nixon proposed a 2-tiered ABM system in 1969; but, after intense Congressional opposition, shifted his stance and actually negotiated the ABM Treaty of 1972 which indefinitely limited missile defenses. President Reagan proposed a radical change in U.S. strategy, urging the adoption of

---

[51] There are many countries with which the U.S. does not have a security treaty but where Americans supply protection (for example, Israel). There are also a large number of countries that benefit without any direct ties, through U.S. guarantees of freedom of the seas.

"population defense." Both of these attempts to change U.S. options and policy were met with fierce opposition. Neither President Nixon nor Reagan was able to get broad-based support for changing American strategy or to deploy missile defenses for the civilian population.

4) Now that the Cold War is over, the threat of large-scale missile attack on the U.S. has been dramatically reduced by two principal factors: (a) political changes in the governance of the Soviet Union / Russia, and (b) the successful ratification of the START I and II treaties. Hence, many of the original assumptions about "Deterrence" are now subject to challenge.

5) "Deterrence" was an acceptable strategy when: (i) both superpowers had roughly comparable resources to devote to defense, and (ii) both sides were willing to accept nuclear parity. When these conditions do not hold, it is dangerous to assume that "Deterrence" will work.

Nuclear Deterrence did not fail at the time of the Gulf War of 1991. However, there is no question that, despite virtually certain defeat, Saddam Hussein refused to heed allied threats and withdraw from Kuwait. If Saddam Hussein had possessed nuclear weapons at that time, there seems little reason to believe that he would have been more worried about "Deterrence" than he was in 1991 when he was playing an even weaker hand.

6) The management and several key policy decisions of the Strategic Defense Initiative under Presidents Reagan and Bush generated so much opposition that there is little Congressional support for NMD. It will not be possible to have BMD deployments unless confidence is established both in the Congress and among the general public regarding the management and operation of BMD systems.

7) "National Defense" (i.e. missile defense of the U.S. homeland) should be seen as essentially comparable to catastrophic health insurance. One hopes never to use it, and under most circumstances it will not be used; however, if it can be afforded, there is a strong argument for coverage.

8) "Deterrence" is an elegant doctrine for a bi-polar world, but, as the character of threats changes, it is simply not adequate to deal with the complexity of the new situation we face.

Criteria for Evaluating BMD

Opponents of NMD vehemently attacked the Reagan SDI using three main arguments: (1) space-based BMD deployments would be a risk to the "ABM Treaty regime,"[52] (2) NMD deployments and possibly even NMD development efforts would produce counter-measures from the USSR,[53] and (3) NMD could be destabilizing both in peacetime and in crisis situations.[54]

These arguments are impossible to accept or refute definitively because they involve hypothetical assumptions about how other nations will react in the future. Nevertheless, as discussed above, all of these contentions were made during the Cold War and at a time when it was plausible that the USSR would react in both a negative and potentially hostile fashion. Yet, whatever the probability of a negative reaction was then, it is far less probable today.

The most concise, comprehensive set of criteria stating how the U.S. State Department intended to evaluate NMD came from Paul Nitze in a speech on February 20, 1985.[55] The *Nitze Criteria* were that strategic or NMD deployments should not be made unless: (1) the system could be demonstrated to be operationally effective against a full-scale Soviet attack, including counter-measures, (2) the BMD must be able to survive direct attack if the system itself was a target as well as sites in the U.S., and (3) the system must be cost-effective at the margin.

Many observers viewed the Nitze Criteria as a "way to kill SDI."[56] No one should argue with having an effective and survivable system. The key criterion was the third and its intent was, presumably, to require that U.S. expenditures on BMD not be foiled by new or expanded offensive systems in the USSR.

There is a key analytic flaw, however, in the third Nitze criterion. There is no reason why one nation which may have a higher national income and more advanced technology should limit itself by requiring that a given investment

---

[52] Carnesale, A., "Managing the ABM Treaty Regime: Issues and Options," in A. Chayes and P. Doty, eds., *Op. Cit.*, pp. 217-238.

[53] See the arguments by R. Garwin cited in Footnote 17 above.

[54] Kincade, W., "The SDI and Arms Control," in *Strategic Defenses and Soviet-American Relations*, S. Wells and R. Litwak, eds., (Cambridge, Mass.: Ballinger, 1987), pp. 101-124.

[55] See, Nitze, P., "SDI, Arms Control, and Stability: Toward a New Synthesis," *U.S. Department of State, Current Policy, No.845*, (Washington, D.C.: U.S. Government Printing Office, June 1986). This was the position of Mr. Nitze and Secretary of State Schultz, but it was not accepted by many on the NSC staff or in the Defense Department. President Reagan appears not to have taken a formal position on this.

[56] Talbott, S., *Master of the Game*, (New York: Vintage Books, 1988), p. 218.

be more efficient than the counter-response. In fact, it may be perfectly rational for a wealthy country to outspend an opponent by a considerable margin if the people, income, and assets being protected are worth more than the investment.

We find examples of this all the time where higher income individuals carry more insurance and, in some cases, pay for private security protection. Thus, the relevant economic criterion is: Does the gain in security outweigh the cost?

To put this in more formal terms, we need to ask two questions:

- What is the probability of the U.S. being involved in a military conflict in the next 20 years? and
- If the U.S. is involved, what is the probability that ballistic missiles will be used against: (a) U.S. troops or military assets deployed overseas? and (b) military or civilian targets in the continental U.S.?

To compute the net present value of a ballistic missile defense for the U.S., the following approximations can be used:

$$PV = \frac{PBMA \times ED \times \frac{BSA}{C}}{(1+r)^n}$$

Where:

PV	=	present value
PBMA	=	probability of missile attack
ED	=	effectiveness of the defense
BSA	=	benefit of stopping the attack
C	=	life cycle cost of the system
r	=	social rate of discount
n	=	number of years system operates

Clearly a present value calculation is a narrow technique based on rigid assumptions, and the pros and cons of this technique will be assessed in Chapter 6. Yet, examples will be shown to illustrate that even low probabilities of hits warrant spending substantial sums on an effective BMD.[57] Thus, the

---

[57] Obviously the humanitarian and social losses of a nuclear attack on an urban nation would be enormous. The only purpose of these calculations is to give the reader an idea of the stock of capital (human and physical) that would be destroyed if an attack were not blocked. For estimates of human and physical capital stock in the

criteria to be used in this volume will be Nitze's first two (effectiveness and survivability) plus a refined economic criterion which weighs the cost of the system against the insurance policy being bought.

The approach suggested here could be criticized as being too static, and it almost certainly would be if there was a major opponent of the U.S. willing to try to counter American moves. Yet, as discussed above, neither Russia nor China is in that position now, so it appears the U.S. could lead in a shift toward defenses with minimal short-term negative responses. As long as the U.S. continues to provide "extended deterrence" for its principal allies in NATO as well as Japan and South Korea, a gradual American shift toward effective TMDs and a limited NMD would be stabilizing and could also encourage a global transition in that direction.

Links of BMD to a Broader National Security Strategy

During the past 30 years, the U.S. has had essentially a three-tiered military strategy: (i) "Deterrence" as the top priority and the overarching means to keep the Soviet Union / Russia from attacking U.S. troops or soil, (ii) conventional forces to deal with non-nuclear conflicts, and (iii) the Single Integrated Operating Plan (SIOP) if nuclear war actually did break out.

Since "Deterrence" will now be far less focused on the Soviet Union and more concerned with third world threats, and the SIOP is an even more remote possibility, how should conventional forces and a new variant of Deterrence be integrated? There are at least three general problems warranting attention, which we will address in depth in Chapter 6.

First, as noted, the spread of technology has led to the spread of military capability and a sharp increase in the number of countries with the capacity to service and use ballistic missiles if they can buy them. There is also a smaller group (India, China, North Korea, and Israel) who have already succeeded at manufacturing ballistic missiles domestically. Therefore, the Missile Technology Control Regime (MTCR) which focuses on sales of missiles and missile parts will need to be broadened to include judgments about "dual-use" industrial technology. This will create the same type of debates that have wracked the member nations of COCOM (the Coordinating Committee

---

(Continued)
U.S., see, J. Kendrick, *The Formation of Stocks of Total Capital*, (New York: National Bureau of Economic Research, 1976) and D. Jorgensen and B. Fraumeni, "Investment in Education and U.S. Economic Growth," *Mimeo.*, (Cambridge, Mass.: Harvard University, 1991).

established to keep key technology from the Communist countries). Yet, it is probably best to recognize this dilemma now and start dealing with it.

Second, as discussed, "Deterrence" between the U.S. and Soviet Union was based on having roughly comparable evaluations of the loss of human life and property. If, in the future, the U.S. finds it is dealing with an antagonist who values the loss of life differently or who seeks to gain a bargaining advantage from threats to the U.S., American policy-makers might feel the need to take preemptive action to forestall the opponent escalating to chemical, biological, or nuclear weapons. This process, in itself, could be destabilizing as more countries saw the U.S. as potentially hostile rather than as a benign "keeper of the peace." Perhaps the U.S. will need to set clearer guidelines about what actions it considers hostile to its interests and to define more precisely the extent of its military commitments.

Third, the U.S. needs to develop an operational doctrine for the use of TMD. The 1972 ABM Treaty prohibits either the U.S. or the Soviet Union from developing mobile ballistic missile defenses, yet this is exactly what TMD systems are. Although current TMD systems can only deal with slower moving intermediate and short-range ballistic missiles, as technology improves, TMD systems will almost certainly be able to defend against certain types of ICBMs. Therefore, the U.S. and USSR must either renegotiate the ABM Treaty to deal with this technological change or gradually see the replacement of "Deterrence" with "Partial Defense" from growing numbers of TMD systems.

*   *   *   *

Now we will turn first to an examination of how U.S. nuclear doctrine changed over time and then to the specific evaluation of TMD and NMD systems.

2

The Changing Meaning of Deterrence

For most of the Cold War period (1949-89), the Soviet Union was the principal challenge and focus of U.S. foreign policy efforts. The chosen American strategy of "containment" was designed to limit the spread of Marxist-Leninist ideology and to thwart the expansion of Soviet influence through territorial gains, alliances, or client-state relationships.[1]

The active military strategy, used to reinforce the goals of containment, had three elements: (1) deploying conventional forces on the periphery of the USSR, (2) supplementing these "forward deployed" ground, naval, and air units with tactical nuclear weapons to be employed if conventional weapons proved inadequate, and (3) maintaining strategic nuclear weapons for targeting against the Soviet homeland if either the conventional or tactical nuclear forces failed.[2] Thus, the American intent was to deter the USSR through a tiered array of potential responses. If there was Soviet aggression and the first tier of defense proved unsuccessful, then carefully defined steps of escalation were available.

There was also a more passive aspect to the strategy that dealt with the

---

[1] See George Kennan's "X" article, "The Sources of Soviet Conduct," *Foreign Affairs*, July 1947, Vol. XXV, No. 4., for the intellectual origins of "containment," but a far more limited statement of objectives than the U.S. actually pursued.

[2] For an elaboration of the links between this force structure and U.S. strategic objectives, see F. Jenkins, "Implications of Defenses and the ABM Treaty for U.S. Strategic Arms Control Policy in the 1990s," *Mimeo.*, (McLean, Va.: S.A.I.C., October 1991).

very unlikely event that the Soviet Union directly attacked the American homeland. Interestingly, much of the literature on security and most of the popular writing about defense planning during the Cold War addressed this possibility.

The elaborate intellectual apparatus of Deterrence and much of our discussion below concerns how the U.S. designed sufficiently redundant strategic forces so that they could absorb a Soviet first strike and still retaliate with a devastating response. Yet, a strategic nuclear exchange was always recognized as a low-probability hypothetical and the forces designed to deal with it rarely consumed more than 20% of the U.S. defense budget. Hence, it is fair to say that American global military strategy had essentially two parts:

- a forward-deployed mix of general-purpose forces, and
- strategic nuclear forces with the dual objective of deterring a first strike and providing the last move in a chain of escalation if a regional war was spreading.

The intent of this chapter is to highlight four broad issues:

1) The *logic* behind Deterrence and an explanation of the major debates that led to the main changes and refinements in the doctrine during the past 30 years;
2) The *links* between Deterrence as a doctrine and the major choices made in U.S. arms control policy;
3) The *manner* in which Deterrence thinking and arms control policy have shaped the U.S. force structure, and how the momentum in these directions has carried into the 1990s when the politico-military environment is strikingly different from the Cold War Era; and
4) The *implications* of the different approaches taken by the Reagan, Bush, and Clinton administrations to BMD and the significance for U.S. strategy *if* the broad directions favored by these respective presidents were actually implemented.

These four areas are obviously interrelated and cannot be completely separated in this fashion. Nevertheless, for the purposes of this discussion, covering them in this order will explain the major doctrinal debates affecting current choices on BMD.

It will be argued below that, despite the Reagan Administration's early (1981-82) statements about prevailing in a protracted nuclear war and the President's clear preference for strategic defense, the U.S. nuclear forces and targeting plans did not change significantly in the 1980s. Moreover, Deterrence remained the actual U.S. strategy despite major efforts by the Reagan Administration to shift the focus from offense to population defense.[3]

Ironically, the Bush Administration, which was so cautious about many of its policy initiatives, actually made a key change in U.S. strategy. By proposing, and getting Congressional acceptance for the near-term deployment of a thin, nationwide NMD, the Bush Administration legitimized the concept of missile defense. The overall Bush concept for BMD was called Global Protection Against Limited Strikes (GPALS).[4]

Although the space-based parts of GPALS are not being continued by the Clinton Administration, the Congress did pass the Missile Defense Act of 1991 which gave the first authorization for NMD in two decades. Also, there is now mainstream acceptance of the usefulness of BMD which was not the case during the bolder and more assertive Reagan years.

The Logic of Deterrence

In the first few years after World War II, there was more progress made on establishing international economic institutions than on security arrangements. Although there was a brief period of enthusiasm about the UN, this faded quickly once the Cold War made the UN Security Council a debating society more than a forum for substantive conflict resolution. There was also an exhaustion from the scale and horrors of a truly global war.

Nevertheless, by mid-1947, two of the most prominent theories examined below had already been published. As discussed above, George Kennan's extraordinary essay, "The Sources of Soviet Conduct," became the essential U.S. rationale for dealing with the USSR, and Bernard Brodie's book, *The Absolute Weapon*, proposed many of the early ideas that ultimately became the foundation for Deterrence.

Brodie argued:

---

[3] President Reagan's March 1983 speech, launching SDI, acknowledged that the shift to population defense might not be until after the year 2000, but that was clearly his goal. See Appendix D.

[4] Many of the top leadership in the U.S. military were unconvinced about the feasibility of missile defense for the general population, but did see GPALS as a means for defending the retaliatory forces. GPALS consisted of three parts: TMDs, a thin NMD from Ground-Based Interceptors, and a global defense from space-based interceptors, Brilliant Pebbles.

> Thus far the chief purpose of our military establishment has been to
> win wars. From now on, its chief purpose must be to avert them.[5]

and

> superiority in numbers of bombs is not in itself a guarantee of super-
> iority in atomic bomb warfare.[6]

These two points not only illustrated that atomic warfare was vastly greater in
its scale of destruction than conventional war, but that it required a different
strategy as well. For purposes of analyzing the evolution of Deterrence doctrine, it is useful to divide the post-World War II era into four periods of dis-
tinctly different U.S. strategy: (a) 1945-52, (b) 1953-60, (c) 1961-73, and
(d) 1974 - present.

1945 - 1952

During the initial optimism about "Collective Security" and the role of the
UN, the U.S. proposed to keep the world non-nuclearized. The "Baruch Plan"
was presented with considerable histrionics:

> We are here to make a choice between the quick and the Dead......If
> we fail, we have damned every man to Fear. Let us not deceive our-
> selves: We must elect World Peace or World Destruction.[7]

The Baruch Plan proposed an international authority, with powers of inspection and control, to prohibit the mining and processing of nuclear materials for
weapons. As an incentive for others to agree to forego development of
nuclear weapons, the U.S. promised to destroy its nuclear systems once the
international authority was established and effectively operating.

The Soviet Union had no intention of giving up the right to develop nuclear
weapons, and the Baruch Plan was quickly dismissed. Various similar plans
have surfaced since then, including one by Thomas Finletter who suggested
that there be a 'Disarmament Executive' composed of minor powers who

---

[5] Brodie, B., *The Absolute Weapon*, (New York: Harcourt Brace, 1946), p. 76.

[6] Brodie, B., *Ibid.*, p. 46.

[7] "The Baruch Plan Statement by the U.S. Representative to the United Nations
Atomic Energy Commission," June 14, 1946, in U.S. Department of State, *Docu-
ments on Disarmament, 1945-1959, Vol.1*, (Washington, D.C.: U.S. Government
Printing Office, 1960), p.7.

would attempt to limit nuclear proliferation and police world conflicts.[8]
Needless to say, these plans were too intrusive to be accepted by both major
and minor powers who were intent on developing a nuclear capability.

In 1947 and 1948 the West Europeans were already troubled by the Soviet
consolidation of control in Eastern Europe; and this helped speed the decision
to ratify the NATO Treaty in July 1949. The U.S. leadership saw NATO as a
way to form a broad military coalition and as a political means to enhance
confidence among the Western nations. This confidence was immediately
tested in August 1949 when the USSR detonated its first nuclear weapon and
it became clear that NATO needed an explicit strategy for deterring Soviet
use of its nuclear capability.

National Security Memorandum # 68, approved in January 1950, went farther than the Containment Policy and called for an explicit build-up of U.S.
military strength to resist potential Soviet challenges. Yet, both the NATO
alliance and the approach taken in NSC-68 were seen as means to bolster
defense if the West was attacked. The West had no plans to initiate a first
strike and was implicitly recognizing that there was going to be some type of
nuclear stalemate with the USSR.

In explaining to Congress the rationale behind NSC-68, Secretary of State
Dean Acheson said:

> The best we can make of our present advantage in retaliatory air
> power is to move ahead under this protective shield to build the
> balanced collective forces in Western Europe that will continue to
> deter aggression after our atomic advantage has been diminished.[9]

This was an important and interesting position for the U.S. Government to
take publicly and it prefigured the debate about the Flexible Response
Strategy in the 1960s. Acheson was acknowledging that, at some point, the
USSR would catch up in nuclear weapons, and the U.S. and NATO would
then need to have adequate conventional forces to deal with the Soviet
military strength.

This led the Truman Administration to invest considerable political capital
in the effort to expand U.S. and West European conventional forces. By
February of 1952, the NATO members agreed in Lisbon to specific targets

---

[8] Finletter, T., *Power and Policy*, (New York: Harcourt Brace, 1954), p. 392.

[9] Acheson, D., Testimony of the Secretary of State, Hearings of the Senate
Foreign Relations Committee, *Assignment of Ground Forces of the United States in
the European Area*, (February 1951).

for expanding both tactical aircraft and army divisions. The expectation was that NATO aircraft would be increased from 4000 to 9000 and the total number of active duty and reserve army divisions would grow from 50 to 96, with all of this to occur by the end of 1954.[10] Clearly, though there was a sense of urgency regarding the magnitude and location of Soviet forces, these NATO force structure targets were unrealistic and the Eisenhower Administration took a very different approach to the defense of Western Europe.

1953 - 1960

As is now being more fully appreciated by historians, despite his military background, President Eisenhower was skeptical of large defense budgets and the Joint Chiefs of Staff's recommendations. Thus, Eisenhower repeatedly searched for ways to trim security expenditures. In October 1953, the National Security Council endorsed a fundamental change in direction, elaborated in its "Basic National Security Policy" paper (NSC 162/2).[11]

Instead of concentrating on building up conventional forces, the Eisenhower strategy was to rely on nuclear weapons as the principal means to deter the Soviet military. The U.S. threat of "Massive Retaliation" was seen as the most effective way to avoid a large scale war, and Europe was seen as the principal arena where events might escalate to direct military conflict.

Thus, the key Eisenhower innovation was to plan on using nuclear weapons in a dual role: (a) as a deterrent, and (b) as a means to limit escalation. This was seen as the most efficient use for nuclear weapons and as a way to limit the need for the costly build-up of conventional forces planned by the Truman Administration.

Despite much of the subsequent public rhetoric, NSC 162/2 did not specifically state that nuclear weapons would be used outside of Europe; and, like the Acheson position, it also anticipated a time when the growth of the Soviet nuclear stockpile would produce a stalemate with the U.S. In fact, President Eisenhower "personally added to the original paper a requirement to reconsider the 'emphasis on the capability of inflicting massive retaliatory damage' if this strategy 'came to work to the disadvantage of national security'".[12]

---

[10] Osgood, R., *NATO, The Entangling Alliance*, (Chicago, Ill.: University of Chicago Press, 1962), pp. 87-88.

[11] For a complete text of NSC 162/2, see *The Pentagon Papers*, Gravel edition, Vol. 1, (Boston, Mass.: Beacon Press, 1971), pp. 412-429.

[12] Freedman, L., *The Evolution of Nuclear Strategy*, (New York: St. Martin's, 1989), p. 83.

Nevertheless, as elements of NSC 162/2 were leaked to the press, the focus was predominantly on "Massive Retaliation" as the new doctrine. This forced the Secretary of State, John Foster Dulles, to clarify what the policy actually was. Dulles wrote an article for *Foreign Affairs* in which he played-down the reliance on strategic bombing, but argued that it was best for aggressors not to know exactly how nuclear weapons would be used.[13]

The initial controversy over the Eisenhower change in doctrine gradually faded; but, among strategic analysts, there was a growing concern that the threat of "Massive Retaliation" and using strategic nuclear weapons to target the Soviet homeland was not credible if there was a limited conflict. There were two very basic problems: (1) even in Europe, conventional forces or tactical nuclear weapons would be a less destructive way to stop what might be a limited war, and (2) outside of Europe, it did not seem reasonable to expect the U.S. to risk global conflict unless its vital interests were threatened.

In 1957 Henry Kissinger published a very influential critique of "Massive Retaliation" as a doctrine. Suggesting that future wars would not be replays of World War I or II and that the objective would be to destroy military units not to hold terrain, he argued in favor of greater reliance on tactical nuclear weapons.[14]

There was some intense criticism of Kissinger's view from those who thought that nuclear weapons could not be used precisely and that their initial use would lead to wide, civilian collateral damage and to a greater risk of escalation.[15] Nevertheless, many other analysts concluded that there was merit in using tactical nuclear weapons in limited wars to deal with concentrations of armor and to avoid immediate decisions to initiate a strategic nuclear exchange.[16]

Although this critique of "Massive Retaliation" had an effect on how quickly the U.S. planned to escalate beyond conventional weapons, it also legitimized the use of tactical nuclear weapons. Thus, throughout the Eisenhower Administration, there was a steady effort to build strategy around

---

[13] Dulles, J.F., "Policy for Security and Peace," *Foreign Affairs*, April 1954, Vol. XXXII, No. 3.

[14] Kissinger, H., *Nuclear Weapons and Foreign Policy*, (New York: Harper, 1957).

[15] See, for example, W. Kaufmann, "The Crisis in Military Affairs," *World Politics*, July 1958, Vol. X, No. 4.

[16] Osgood, R., *Limited War: The Challenge to American Strategy*, (Chicago, Ill.: University of Chicago Press, 1957).

nuclear systems and to convince the European members of NATO that this was wise policy.

> This campaign (for a nuclear strategy) began in earnest in late 1953 and mounted steadily from 1954 to 1956. It culminated at the NATO summit of fall 1957, when the alliance formally adopted a new, nuclear-oriented military strategy (MC 14/2) and approved important programmatic decisions to field the nuclear forces required by it. Eisenhower's last three years were largely devoted to implementing these decisions and fending off criticisms directed at them, while trying to strike a balance between the allies' demands for greater sharing of control over nuclear weapons and congressional resistance to provide it.[17]

Thus, the thrust of NATO's MC 14/2 undercut the Lisbon agreement and the plans to build up conventional forces. Hence, the Eisenhower approach created a new turn in the strategic debate, by: (1) initiating the argument over "Extended Deterrence" (i.e. would the U.S. really keep commitments to its allies and use nuclear weapons if the U.S. homeland was not directly threatened?), and (2) entering the second phase of the seemingly never-ending argument over selecting the right mix of nuclear and conventional forces.

1961 - 1973

The Kennedy Administration marked still another turning point in strategic thinking because it pursued the apparently contradictory goals of a major arms build-up simultaneously with launching significant efforts at achieving arms control agreements with the Soviet Union. This reflected some of the newest theories at the time on the relationship between arms control and strategy. In their highly influential 1961 book, Thomas Schelling and Morton Halperin said:

> We believe that arms control is a promising, but still only dimly per-
> ceived, enlargement of the scope of our military strategy. It rests
> essentially on the recognition that our military relation with potential
> enemies is not one of pure conflict and opposition, but involves
> strong elements of mutual interest in the avoidance of a war that nei-

---

[17] Kugler, R., *Commitment to Purpose: How Alliance Partnership Won the Cold War*, (Santa Monica, Cal.: RAND, 1993), p. 76.

ther side wants, in minimizing the costs and risks of arms competition, and in curtailing the scope and violence of war in the event it occurs.[18]

Robert McNamara was Secretary of Defense from 1961 to 1968, and, in that period, there was an extraordinary burst of activity in strategic thinking and in efforts to manage the U.S. Government's largest bureaucracy. Several major changes were made in nuclear targeting doctrine, three important arms control agreements were signed,[19] the basis was laid for the Strategic Arms Limitation talks (which led to SALT I), and a new doctrine was developed for the mix of NATO's nuclear and conventional options for defending Europe.

This sweeping set of McNamara initiatives generated resistance from the U.S. career military, from the Congress, from many defense contractors, and from the NATO allies.[20] Nevertheless, it is fair to say that, by 1968, the basic strategy used today for U.S. nuclear and conventional forces was already in place. There have been some critical changes since, particularly in nuclear targeting doctrine and in the goals for arms control negotiations which we will discuss below. However, the overarching intellectual construct of the McNamara years (Deterrence and Flexible Response) are still preeminent.

There were two essential objectives that McNamara set for strategy during the Kennedy Administration and they carried over through the Johnson years: (a) centralizing control of decision-making, and (b) attempting to facilitate stability during a crisis by broadening the range of military options available to the President or National Command Authority.

President Eisenhower had preferred the threat of massive nuclear retaliation because he was convinced that was the most efficient way to keep any opponent from attacking the U.S. or its allies. The Kissinger focus on tactical nuclear weapons was seen as an intermediate option that could be militarily effective without taking the potentially catastrophic step of escalating to strategic nuclear weapons.

McNamara thought the U.S. should have a full panoply of options, incorporating a broad range of technologies and systems but putting the initial reliance on conventional weapons. This meant seeing nuclear weapons as a

---

[18] Schelling, T. and M. Halperin, *Strategy and Arms Control*, (New York: 20th Century Fund, 1961), p. 1.

[19] The Outer Space Treaty (1963), the Limited Test Ban Treaty (1963), and the Non-Proliferation Treaty (1968).

[20] For a critical, but detailed summary of the reactions to McNamara, see D. Shapley, *Power and Promise*, (Boston: Little-Brown, 1993), pp. 269-288.

back-up rather than the first choice.[21] Although this generally pleased
strategists who had long been worried about the risks of the Eisenhower
approach, there was a high political cost because it meant going back to elements of the Lisbon Declaration and trying to buildup conventional forces.
This was not only expensive but unpopular in Europe because many of the
NATO governments saw "conventional options" as a way for the U.S. to
"decouple" itself from firm commitments to defend Europe.

The first principal McNamara venture in nuclear strategy was popularly
labeled "City Avoidance." At the June 1962 University of Michigan commencement, the Secretary of Defense said:

> ....principal military objectives, in the event of a nuclear war stem-
> ming from a major attack on the (NATO) alliance, should be the
> destruction of the enemy's military forces, not his civilian
> population.[22]

Thus, City Avoidance was an effort to convince the Soviet leadership that
the U.S. would not initially aim at cities and that the USSR should design its
forces to concentrate on military targets.

This seemed like a laudable objective, but it immediately ran into a number of problems. First, many U.S. military bases were close to cities and
there would be collateral damage even if the Soviet Rocket Forces intended
only to hit the bases. Second, the USSR was having trouble with the accuracy
of its missiles; so, in an attack, it might be hard for U.S. officials to know
what the intended target was. Third, and most important from the standpoint
of crisis stability, if the U.S. initiated the use of nuclear weapons against
Soviet military bases or field units, it might be difficult for the Soviet leadership to know if the attack was the beginning of a "first strike" against their
secure missiles or just a selective "counter-force" response to conventional
losses elsewhere.

Curiously, although the Cuban Missile Crisis in October 1962 was seen as
a brilliant success at nuclear diplomacy, it undercut elements of McNamara's
argument. Subsequently it became clear that the USSR had backed down
because the U.S. had an overwhelming superiority in both numbers of missiles

---

[21] See W. Kaufmann, *The McNamara Strategy*, (New York: Harper and Row,
1964), pp. 74-75, for a discussion of the relationship between conventional and
nuclear systems.

[22] McNamara, R., "Defense Arrangements of the North Atlantic Community,"
Department of State Bulletin, July 9, 1962, Vol. 47, p. 68.

and total destructive power (megatonnage). Moreover, President Kennedy had threatened a massive attack, not a precise counter-force response if the Soviet Union breached the embargo around Cuba. Also, in a little noticed move, the President had authorized dispersing American B-47 bombers to civilian airfields. In essence, this meant that the U.S. itself was not making a sharp distinction between military and civilian targets.[23] It also became clear that the U.S. had local conventional superiority and thus superiority at every level of escalation.

As the concept of City Avoidance (and its inherent emphasis on counter-force targeting) developed more critics, McNamara shifted his emphasis to "Assured Destruction" as a means to provide a theoretical foundation for choosing the size and objectives of the U.S. missile force. This put American strategic planners in the ghoulish position of estimating the level of carnage and destruction that was necessary to deter a Soviet attack. McNamara thought that, if the U.S. had an assured second-strike capability of being able to annihilate 20% of the Soviet population and 50% of its industrial capacity, that should be enough to deter any thinking of launching an initial attack on the U.S.[24]

The element of the McNamara approach that many specialists and the public had the greatest trouble accepting, however, was the argument for mutual vulnerability with an opponent. As cited above, even as early as 1951, Secretary of State Acheson had recognized that a determined Soviet Union could eventually reach parity with the U.S. or at least close the gap in strategic systems. McNamara's boldness was in his willingness to state this openly and actually build a policy around it. Thus, "Mutual Assured Destruction (MAD)," included in the term "Deterrence" in this volume, became the centerpiece of U.S. nuclear policy in the 1960s. It was based on the assumption that missile defenses were not cost-effective and that mutual vulnerability would create crisis stability.

Much of the intellectual framework for MAD came from game theory and simulations done at the RAND Corporation in the 1950s.[25] Game theory

---

[23] Quester, G., *Nuclear Diplomacy: The First Twenty-Five Years*, (New York: Dunellen, 1970), p. 246.

[24] Enthoven, A., and K.W. Smith, *How Much Is Enough? Shaping the Defense Program 1961-69*, (New York: Harper & Row, 1971), p. 174.

[25] See, F. Kaplan, *The Wizards of Armageddon*, (New York: Simon and Schuster, 1983), pp. 66-71, for a discussion of the RAND usage of game theory. For an overview of the application of game theory to defense issues (including deterrence, arms races, and missile defense), see, S. Brams and D. Marc Kilgour, *Game Theory and National Security*, (New York: Basil Blackwell, 1988).

provided a construct for analyzing decision-making when an opponent's response is uncertain. Thus, it was seen as a rigorous way to analyze potential moves by the Soviet leadership when little was known about the internal maneuvering at the Kremlin.[26] Also, Air Force analyses showed that, per dollar allocated, defenses (air defense, NMD and civil defense) could not keep up with the Soviet ability to add offenses.

Lawrence Freedman concisely summarizes an era in U.S. strategic thinking:

> The underlying assumptions of the Theory of Mutual Assured Destruction were that, for the foreseeable future, the offense would be able to maintain an advantage over the defense. Because of this, all that one could do to prevent the other from inflicting crippling devastation was to threaten retaliation. The lesson drawn from this assumption for the purposes of force planning was that one need only ensure a sufficiency of offensive forces to assure destruction after allowing for all feasible improvements in the first-strike capability of the other side. The lesson drawn for arms control was that, as every improvement in one side's defense provided no extra security but merely a spurt to the offense of the other, once both sides ceased making defensive moves, forces could stabilize at current levels. The lesson for crisis management was the growing uselessness of nuclear weapons for purposes other than deterring the nuclear weapons of the other side.[27]

In the 1960s, the intellectual fascination was with MAD. Yet, the tough foreign policy choices dealt with how to defend Western Europe and the growing morass of the Vietnam War. These issues were all related in ways that did not fit the pristine, analytical construct advocated in the McNamara pentagon.

Not only did the French distrust U.S. intentions, leading to their withdrawal from NATO's integrated command, but the Germans were concerned that the U.S. might be losing sight of how to actually defend key parts of West

---

[26]   There have been considerable innovations in game theory since the 1950s, with the development of techniques to model dynamic processes and look several moves ahead before choosing a particular strategy. For an exposition of these methods, see, S. Brams, "Theory of Moves," *American Scientist*, Nov.-Dec. 1993, Vol. 81, pp. 562-570.

[27]   Freedman, L., *The Evolution of Nuclear Strategy*, (New York: St. Martin's Press, 1989), p. 259.

German territory. McNamara's focus on enhancing conventional forces and creating greater options for the American President sounded to many Germans like a process which could lead to indecision if NATO's forces were being overrun by the Warsaw Pact.

In addition, the West German parties on the political Left were beginning to advocate "Ostpolitik" and favoring cooperation rather than confrontation with the Marxist-Leninist regimes in Eastern Europe and the USSR. Although truly impressive conventional forces would have, as theorized, delayed the time after an attack before nuclear weapons would be used, many West Germans saw this as an invitation to use their country as a battlefield.

Not surprisingly, most West Germans wanted NATO committed to defending the inter-German border, and were opposed to "defense in depth" or other strategies that would have let Warsaw Pact troops move deep into West Germany and then have NATO attack their lines of communication and less-concentrated armor. Additionally, many European commentators were convinced that the U.S. was bogged down in an unwinnable war in Vietnam; and that this would not only divert attention from NATO's needs but potentially exhaust U.S. patience with overseas commitments as well.

Hence, a very elaborate brokerage developed. After moving the NATO command headquarters out of France into Belgium, there was a broad recognition that Germany's role in Western force planning had been substantially strengthened. Since the U.S. was tied down in Southeast Asia and at the same time was asking for increased West European defense expenditures, this created a perfect opportunity for the Germans to bargain for more influence on NATO strategy.

In June of 1966, the NATO member countries agreed to have the Belgian Foreign Minister, Harmel, conduct a working group on the relationship between East-West issues and military plans. Simultaneously, the Military Committee was authorized to reevaluate the assumptions and tactics incorporated in NATO's 1957 military plan, MC 14/2.

The military dilemma that NATO faced mirrored the strategic nuclear debate that led to acceptance of mutual vulnerability. In 1966 NATO commanders had over 7000 nuclear warheads that could be used for theater operations. The Western assumptions were that these tactical nuclear weapons (available as bombs, artillery shells, or on short range missiles) would be used to break up the growing concentrations of Soviet mechanized infantry and armour. Yet, Soviet tactical nuclear systems (especially on FROG and SCUD missiles) were being rapidly deployed, and it appeared to be only a matter of time until the Warsaw Pact tactical systems roughly equaled NATO's. Moreover, even the use of tactical nuclear weapons would have produced vast collateral damage.

The problem was that the Soviet Union continued to spend very high percentages of its GNP on defense[28] and gave every indication that its leadership was willing to spend what it took to continue to field the 80-100 divisions of the Warsaw Pact.[29] If the Soviet Union was gradually reaching both strategic and tactical nuclear parity and had a predominance in conventional forces, then, eventually, NATO and the U.S. could not plausibly threaten to escalate if a conflict broke out.

This placed a very high premium on figuring out ways to thwart the power of Warsaw Pact conventional forces. Ultimately, it led to innovations in systems and tactics that, on paper, looked promising as a way to blunt the advantages of the more than 2:1 numerical ratio of Eastern to Western conventional forces. Nevertheless, in the short-run, the NATO governments had fewer options and agreed to expand conventional forces to avoid being coerced by the Warsaw Pact.

Thus, in December 1967, after elaborate bargaining, the NATO governments agreed to: (1) establish a five year defense plan with specific targets for improving the Western force structure, (2) accept the Harmel Report which advocated pursuing improved relations with the USSR while simultaneously expanding NATO's military capability, and (3) approve a new military doctrine, MC 14/3, which became widely known to the general public as the "Flexible Response" strategy.[30]

The Flexible Response doctrine necessitated significantly expanded NATO conventional forces but kept the German preference for tactical nuclear systems as an integral part of the military plan. It also took on added obligations by committing NATO to forward defense of the border between East and West Germany, but excluded cross-border operations by NATO. This complicated NATO's defense options somewhat because deployments up close to the border were more exposed to potential attack, and thus needed to be hardened.

---

[28] The exact percentages were hard to estimate at the time because of price distortions in Soviet GNP accounting, but it now appears that the USSR frequently spent over 20% of its GNP on defense (more than three times the U.S. level in the 1960s).

[29] See T. Wolfe, *Soviet Power and Europe, 1945-1970*, (Baltimore, Md.: Johns Hopkins Press, 1970) for a full discussion of Soviet military statements at the time and Western interpretations of their significance.

[30] See R. Kugler, *Commitment to Purpose - How Alliance Partnership Won the Cold War*, (Santa Monica, Cal.: RAND, 1993), pp. 171-190, for a detailed assessment of the content and implications of MC 14/3.

Ironically, Flexible Response reinforced the well-known "Tripwire" concept because it was clear that, if U.S. and NATO troops were overrun, the U.S. would need to become fully engaged. Nevertheless, this proved important for the political consensus inside West Germany to maintain support for NATO and expand defense expenditures. The five year NATO defense program also had an important effect because it set clear deployment and force modernization goals and created group pressure to meet commitments.

The arms control parts of the McNamara construct will be covered below. However, it is interesting to note that the Nixon / Kissinger successes of 1972 (the Strategic Arms Limitation Agreement, SALT I, and the Anti-Ballistic Missile Treaty) were based on the intellectual assumptions of Mutual Assured Destruction. The key Nixon / Kissinger innovation was to recognize that the tension between the USSR and China created the possibility of triangular diplomacy and, thus, the means for extracting greater concessions from the Soviet Union.

1974 - Present

During the Ford and Carter years there were two important moves away from pure Deterrence as envisaged with MAD. In 1974 Secretary of Defense James Schlesinger announced National Security Decision Memorandum (NSDM) 242 which explicitly endorsed nuclear counter-force targeting and limited nuclear options. The goal was to have a wide range of Soviet military targets so that a U.S. President could escalate gradually but precisely, depending upon the progress and scale of the conflict elsewhere. This was to avoid putting the President in the position of having to use nuclear weapons against a Soviet or Eastern European city if the real objective was to stop a Warsaw Pact advance to the West.

As discussed above, elements of this idea were floated by Secretary McNamara as City Avoidance, but ultimately discarded in the 1960s as unworkable because it was felt that using nuclear weapons at all would almost certainly initiate broader strategic exchanges against the U.S. and Soviet homeland. Schlesinger, who had started his career at RAND, however, was clearly more influenced by the Wohlstetter tradition[31] and thought it

---

[31] For an updated variant of this view, see A. Wohlstetter, "Between an Unfree World and None - Increasing Our Choices," *Foreign Affairs*, Summer 1985, Vol. 63, No. 5, pp. 962-994.

was necessary to have nuclear options that demonstrated NATO's will-power and directly reduced the Warsaw Pact's ability to continue its fighting.

The U.S. did not try to develop a missile force that was capable of a disabling first strike against Soviet ICBMs. Going for that type of offensive capability clearly would have been intensely threatening to the Soviet leadership and destabilizing. What NSDM 242 did was let the Soviet military understand that its long-term war-fighting capacity would be at risk early in a conflict, and that the use of nuclear systems would not be limited to tactical weapons at the forward edge of the battle area (as envisaged in NATO's 1967 plan, MC 14/3).

The Carter Administration entered office pledging to cut the U.S. defense budget by 5% in real terms. Yet, the Carter planners quickly found the Soviet Union rebuffing Secretary of State Vance's proposal for deep arms cuts and proceeding rapidly with its deployment of heavy ICBMs (SS-18s) and highly accurate Intermediate Range Ballistic Missiles (IRBMs), the SS-20s.

Although the Carter Administration continued the SALT talks and initialed the SALT II accords in June 1979, Soviet global deployments were so troubling that there was little chance of getting Senate ratification. Then, in December of 1979, when the USSR invaded Afghanistan, the Carter Administration was under considerable pressure to take a stronger stance toward Moscow. In addition to indefinitely postponing the SALT II treaty, Secretary of Defense Harold Brown reassessed strategic force planning.

This led to firm, public commitments to proceed with the M-X (an ICBM capable of carrying 10 independently targeted warheads), various stealth aircraft programs, and a revised strategic nuclear policy. In June 1980, during his last six months in office, Carter signed Presidential Decision (PD) 59 which extended the list of counter-force targets approved in NSDM 242. Despite its earlier statements favoring a purer form of Deterrence, the Carter Administration ended its term saying the U.S. needed a "Countervailing Strategy"[32] to definitively threaten escalation if a war broke out.

If NSDM 242 and PD 59 were incremental but significant moves away from straight Deterrence, neither was seen as adequate by the conservative critics of Carter[33] or among the new leadership in the Pentagon during the early Reagan Administration.[34]

---

[32]   For a description of this approach, see W. Slocombe, "The Countervailing Strategy," *International Security*, Summer 1981, Vol. 6, No. 1.

[33]   For an assessment of the positions taken by the Committee on the Present Danger, one of the most influential groups criticizing Carter, see S. Talbott, *The Master of the Game*, (New York: Vintage, 1988), p. 287.

[34]   For an overview of the Reagan Administration's proposed changes, see, C.

The whirlwind of activity in the early Reagan years was based on a fear that the Soviet Union was actually trying to achieve strategic dominance and that McNamara's hopes for mutual vulnerability through Deterrence were illusory. Two types of evidence were troubling: (1) the deployments of highly accurate ICBMs (SS-18s and SS-19s) which could potentially be used to attack U.S. land-based ICBM silos and unclassified Soviet statements on strategy implying that the USSR saw nuclear weapons as a means for war-fighting not just deterrence;[35] and (2) the consistent maintenance and modernization of air defenses and the anti-ballistic missile system around Moscow. At the same time, Soviet national security specialists published articles advocating active defense which rejected concepts of mutual vul-nerability long favored by supporters of Deterrence.[36] (See Appendix B for a detailed chronology of Soviet BMD efforts).

The Reagan Administration was determined to make a sharp break from the cautious approach taken under Carter. In 1981, the new administration proceeded on two tracks with the USSR: moving aggressively to modernize U.S. strategic forces, while simultaneously saying the SALT process had been a failure and taking the position that arms control negotiations were worthwhile only if they would lead to significant arms cuts. This led to a change in the acronym for the negotiations as they were rechristened the "Strategic Arms Reduction Talks" (START).

The modernization of the U.S. strategic forces was widespread and could not fail to catch the attention of the Soviet leadership. Reagan authorized, and the Congress approved: 100 B-1 B bombers, rapid development of the B-2 Stealth bomber, D-5 warheads to go on the Trident Submarine Launched Ballistic Missiles (SLBMs), procurement of 100 M-X ICBMs, accelerated test-ing of nuclear weapons, and hardening (through the use of fiber optics and other means) of the command, control, and communications systems that would keep the President in touch with his commanders during a conflict.

There was also a very obvious leak of the "Defense Guidance" in 1982. The Defense Guidance is supposedly a highly classified document giving

---

(Continued)

Fred Ikle, "The Reagan Defense Program: A Focus on the Strategic Imperatives," *Strategic Review*, Spring 1982.

[35] For the views of an influential analyst who felt the West was naively misinterpreting Soviet objectives, see, C. Gray, *Nuclear Strategy and National Style*, (Lanham, Md.: Hamilton Books, 1986).

[36] See D. Yost, *Soviet Ballistic Missile Defense and the Western Alliance*, (Cambridge, Mass.: Harvard University Press, 1988) for a review of Soviet programs and public statements regarding ABM systems.

basic guidelines from the civilian authorities in the Pentagon to the military planners. The exceptionally blunt language and objectives stated in the Weinberger guidance were clearly intended to reverse the perceived softness of the Carter period. Instead of "Deterring" or "Countervailing," the Reagan approach was to "Prevail." Several excerpts from the *Washington Post* illustrate how far the U.S. had moved away from pure Deterrence in its stated policy:

> Should deterrence fail and strategic nuclear war with the USSR occur, the U.S. must prevail and be able to force the Soviet Union to seek earliest termination on terms favorable to the U.S.
> The U.S. must have plans that assure U.S. strategic nuclear forces can render ineffective the total Soviet military and political power structure.... so that they have a strong incentive to seek conflict termination short of an all-out attack on our cities and economic assets.[37]

The Reagan agenda was vigorously criticized by many American specialists on the USSR,[38] by long time advocates of pure Deterrence,[39] by specialists on the command and control system who were not convinced that nuclear weapons could be used in such a precise fashion,[40] and by a large number of "peace" and liberal political groups as well. In addition, there were many articles claiming that the proposed Reagan defense build-up would seriously hurt the American economy.[41]

It also turned out that even some very conservative U.S. Senators did not want the M-X missiles deployed in their states. This forced President Reagan and Secretary Weinberger to seek a way to broaden the support for their program. They chose General Brent Scowcroft, who had been the National Security Adviser in the last years of the Ford Administration (and a close confidant of Henry Kissinger, an advocate of the SALT I Treaty), to lead a commission to study options for the strategic forces.

---

[37] Halloran, R., *The Washington Post*, November 10, 1982, p. 3.

[38] See, S. Bialer and J. Afferica, "Reagan and Russia," *Foreign Affairs*, Winter 1982/83, Vol. 61, No. 2., pp. 249-271.

[39] Bundy, McG., "The Bishops and the Bomb," *The New York Review of Books*, June 16, 1983.

[40] Bracken, P., *The Command and Control of Nuclear Weapons*, (New Haven: Yale University Press, 1983).

[41] See, for example, L. Thurow, "How to Wreck the Economy," *The New York Review of Books*, May 14, 1981, pp. 3-8.

Not surprisingly, the Scowcroft Commission concluded that elements of the early Reagan defense program could be trimmed back. The group argued that, even if land-based ICBMs were vulnerable to attack from the SS-18s, U.S. bombers and SLBMs were highly survivable, and that two legs of the "Strategic Triad" were probably enough to ensure deterrence.[42]

The Scowcroft Commission also suggested a rethinking of the concept of deterrence and urged moving away from Multiple Independently Targeted Vehicles (MIRVs) back to the 1960s type deployments where each missile had only one warhead. The Commission specifically advocated the small, mobile missile (Midgetman) as a way to retain deterrent capability without giving an opponent the incentive to attack and conceivably knock out 10 or more warheads per missile.[43] The group never did come up with a better deployment plan for the M-X; and, instead, just suggested that the M-X should be put in hardened Minuteman silos. Nevertheless, the thrust of the Commission was clear, stressing deterrence and casting doubt on "prevailing" in a nuclear conflict.

The low-keyed and pragmatic stance of the Scowcroft Commission report was upstaged, however, by President Reagan's speech one month earlier (in March 1983) advocating an expanded and highly visible program for ballistic missile defense.[44] The plan for a Strategic Defense Initiative was immediately challenged on scientific grounds,[45] but many commentators missed the more important strategic issue, which was the implicit criticism of deterrence in the speech. The President's words were: "Wouldn't it be better to save lives than to avenge them?" This relegated the concept of Deterrence, as enshrined for the previous twenty years, to being a second best concept.

The reaction of those who had worked to establish Deterrence and who were convinced that it was still the best means to assure stability was swift and highly visible. A blizzard of op-ed pieces, articles, and books appeared criticizing the Strategic Defense Initiative (SDI),[46] and some of the most

---

[42] See, *Report of the President's Commission on Strategic Forces*, (Washington, D.C.: U.S. Government Printing Office, April 1983).

[43] Although the USSR had already deployed a number of mobile missiles, the prototype of the U.S. Midgetman proved to have several important technical problems and was never authorized for production.

[44] For a full text of the part of the speech that dealt with the Strategic Defense Initiative, see Appendix D of this volume.

[45] See, for example, Tirman, J., ed., *The Fallacy of Star Wars*, (New York: Vintage Books / Union of Concerned Scientists, 1984).

[46] For example, see the book by *New York Times* writers critiquing SDI: P. Boffey, W. Broad, L. Gelb, C. Mohr, and H. Noble, *Claiming the Heavens -*

influential figures in the American foreign policy establishment began a renewed onslaught against ballistic missile defenses.[47] In many ways this controversy was similar in tone and content to the debate over the proposed Safeguard BMD system in 1969-70.

Although President Reagan was attempting to make a fundamental change in U.S. strategic doctrine, even the true believers in SDI recognized that effective BMD might be a decade or more away. Thus, it is important to note that the Strategic Integrated Operating Plan (SIOP), the actual wars plans with specific nuclear targets identified, did not change dramatically during the Reagan years.

Therefore, if there had been a major East-West conflict during Reagan's term, there was no BMD capable of dealing with ICBMs; and the war would, presumably, have been fought using counter-force concepts similar to those urged by Secretary Schlesinger in 1974. Moreover, as we will discuss below, even during the Bush years (when it was Administration policy to proceed with deploying a space-based missile defense system - Brilliant Pebbles), official statements still stressed "enhancing deterrence" rather than shifting U.S. plans completely to a defense-oriented strategy.[48]

The Links Betv.een Deterrence and Arms Control

In the United States since the mid-1960s, there has been very close cooperation between military specialists who support Deterrence and groups favoring arms control. We have already reviewed the myriad ways in which Deterrence has been refined and modified in the past 30 years. Yet, it is important to make clear that, just as there are numerous variations of deterrence concepts, there are a vast array of approaches to arms control.

(Continued)
Complete Guide to the Star Wars Debate, (New York: Times Books, 1988).

[47] One of the most cited challenges to SDI was by McGeorge Bundy, George Kennan, Robert McNamara, and Gerard Smith in, "The President's Choice: Star Wars or Arms Control," *Foreign Affairs*, Winter 1984/85, Vol. 63, No. 2, pp. 264-278.

[48] The debate need not have been limited to just "Deterrence vs. BMD." Robert Art, for example, was skeptical of minimal deterrence, war-winning nuclear strategies, and SDI. Thus, he proposed deterrence plus some additional threat capability to limit escalation in, "Between Assured Destruction and Nuclear Victory: The Case for the MAD-Plus Posture," in R. Hardin, et. al., eds., *Nuclear Deterrence - Ethics and Strategy*, (Chicago, Ill.: University of Chicago Press, 1985), pp. 121-140.

Rhetorical enthusiasm for arms control has long been popular. In 1930, President Hoover said:

> ...the question before us now is not whether we shall have a treaty with either three or more eight inch cruisers or four less 6-inch cruisers or whether we shall have a larger reduction in tonnage. It is a question of whether we shall move strongly towards limitation in naval arms or whether we shall have no limitation or reduction and shall enter upon a disastrous period of competitive armament.[49]

Despite President Hoover's support and the considerable effort devoted to negotiating the London Naval Agreements of 1922 and 1936, there is now a fair consensus among historians that these efforts were neither particularly effective in their own right nor significant in altering the drift toward World War II.[50] It is certainly not valid to judge all arms control efforts by the inter-war naval experience. On the other hand, it is legitimate to ask: What are the assumptions behind arms control efforts? and What are they likely to accomplish?

Robert Jervis has written succinctly, "If the main objective of arms control is to make war less likely, then any theory of arms control must rest on a theory of the causes of war."[51] Schelling and Halperin do not present causal theories of war, but say that, at any one point in time, arms control may reduce the chances of war by: (1) decreasing the incentive for pre-emptive attack, (2) limiting the incentive for premeditated attack, and (3) lowering the danger of accidental war.[52]

These are strong and plausible propositions, but, obviously, need to be tested against empirical evidence. Much of the debate in the 1970s and 1980s about the effectiveness of U.S. arms control negotiations with the USSR focused on whether these goals had been met.[53]

---

[49] As cited by C. Hall, *Britain, America and Arms Control, 1921-1937*, (New York: St. Martin's, 1987), p. 107.

[50] See, for example, D. Brennan, "The Setting and Goals of Arms Control," in D. Brennan, ed., *Arms Control, Disarmament, and National Security*, (New York: Brazilier, 1961) and R. Rotberg and K. Rabb, eds., *The Origin and Prevention of Major Wars*, (Cambridge, U.K.: Cambridge University Press, 1989).

[51] Jervis, R., "Arms Control, Stability, and Causes of War," *Political Science Quarterly*, 1993, Vol. 108, No. 2, pp. 239-253.

[52] Schelling, T. and M. Halperin, *Strategy and Arms Control*, (New York: 20th Century Fund, 1961), pp. 9-14.

[53] For a detailed critique of the SALT II negotiating process, see, P. Nitze, *From Hiroshima to Glasnost*, (New York: Grove, Weidenfeld, 1989), pp. 359-365.

From the time of the Glassboro Summit in 1967 until the Reagan Administration entered office in 1981, U.S. arms control efforts were essentially based on the following four propositions: (i) both sides would suffer grievous damage in a nuclear war even if one prevailed, (ii) arms control can proceed even between hostile countries, (iii) mutual vulnerability leads to stability, and (iv) appropriately designed mutual vulnerability reduces the incentives for arms races because increasing offensive capability does not necessarily lead to improved ability to coerce an opponent.

Hence, the link between arms control and Deterrence was both intellectual and pragmatic. Bluntly put: if there was no assurance you could stop a missile attack and your opponent was willing to continue to invest in upgrading that capability, then why not threaten him using the lowest cost alternative? Basically, it was hoped that the arms control process would smooth, slightly constrain, and ameliorate the dangerous aspects of superpower competition. The theoretical assumptions above and this logic seemed compelling until analysts saw patterns in Soviet behavior which indicated efforts to seek dominance rather than accept mutual vulnerability.

First, as discussed earlier in detail, though arms control was supposed to convince antagonists of the futility of further investments in offensive weaponry, after SALT I, the USSR proceeded to deploy precisely those types of systems (i.e. SS-18s) which seemed designed to give them an advantage in coercive ability.

Second, although President Carter's decision to cancel the B-1 bomber and go slow on the Trident and M-X programs was not required by SALT I, hopes of changed Soviet behavior did lead to delays in the modernization of the U.S. strategic delivery systems.

Third, despite written agreements to the contrary, the USSR continued to encrypt data sent back during missile launches and reentry vehicle tests. This became a hotly contested issue in 1979 after the Shah of Iran was overthrown and the U.S. lost its ability to closely monitor tests from the Sary Shagan site in Kazakhstan.

These troubling empirical findings led to a dramatic loss of confidence in arms control just as the SALT II treaty was being reviewed by the U.S. Senate. The Senate Foreign Relations Committee voted favorably on the Treaty (by a tally of 9 to 6), but it attached twenty specific conditions for modifications; and the Senate Armed Services Committee actually voted 10 to 7 against the Treaty.

Thus, despite the optimism about strategic arms control generated in the 1960s[54] and uncertainty about what force balances would have been without arms control, there was deep pessimism about the effort by the end of 1970s. Not only did this lead to the Reagan Administration conviction that only arms reductions were worth negotiating over, but it also led to some more fundamental critiques of the process.

Ken Adelman has stated that arms control negotiations consume vast amounts of time of top policy-makers "with little or no relevance to keeping the peace or strengthening the nation."[55] Colin Gray posits an even more pessimistic view, arguing that arms control is a paradox where aggressive nations wishing to change the power balance will not agree to or adhere to covenants that seriously restrict their behavior.[56]

These skeptical views of the likely efficacy of strategic arms control were certainly strengthened by recent evidence of compliance failures in the Iraqi and North Korean violations of the Non-Proliferation Treaty and the marginal effectiveness of the Missile Technology Control Regime.[57] A related and growing issue of concern is whether nations that conform to arms control agreements run the risk of having their readiness hindered while non-complying states "break out" and gain an advantage.[58]

The end of the Cold War has meant that, currently, there is no nation directly threatening U.S. national survival. Therefore, debates about the pros and cons of mutual vulnerability appear abstract. Yet, if the periods before World War I and II are any indication of the future, strategic balances can change very rapidly. Americans now have a fortunate respite when we can speculate on what mix of deterrence and defense is truly desirable. As we will see in Chapters 3 and 4, the nature of the security problems the U.S. now faces is more regional and disaggregated than it was in the 1940s - 1980s. So, it will certainly be harder to draw broad generalizations about optimal security policies, and defense choices are likely to be even more strongly influenced by regional foreign policy objectives.

(Continued)

[54] Confidence that arms control would work and that opponents of it were irrational is illustrated by Paul Warnke's article, "Apes on a Treadmill," *Foreign Policy*, Spring 1975, No. 18, pp. 12-29.

[55] Adelman, K., *The Great Universal Embrace: Arms Control - A Skeptic's Account*, (New York: Simon and Schuster, 1989), p. 34.

[56] Gray, C., *House of Cards - Why Arms Control Must Fail*, (Ithaca, N.Y.: Cornell University Press, 1992), pp. 17-22.

[57] See, J. Nolan and A. Wheelon, "Third World Ballistic Missiles," *Scientific American*, August 1990, p. 34.

[58] For cases on this issue from the 1930s, see C. Fairbanks, Jr. and A. Shulsky, "From 'Arms Control' to Arms Reductions: The Historical Experience," *Washington Quarterly*, Summer 1987, Vol. 10, pp. 59-73.

The Impact of Deterrence and Arms Control Policy
on the U.S. Force Structure

In the 1950s, the strategy of Massive Retaliation shaped the U.S. force structure very decisively. As covered above, President Eisenhower favored cutting the military budget from its Korean War level and saw nuclear weapons as simply a cost effective way to deter an aggressive Soviet Union. The Radford Plan in Europe explicitly substituted nuclear systems for manpower, as a way to deal with the large concentrations of Warsaw Pact infantry and armour in Eastern Europe.

Two challenges to the Eisenhower view ended up making basic changes in the U.S. force structure. As analyzed, Kissinger's *Nuclear Weapons and Foreign Policy* made the case for tactical nuclear weapons, and it was accepted for both military and political reasons. Maxwell Taylor's more technical argument for increased troop mobility and maneuver[59] was controversial and more expensive; but it played a key part in building the rationale for the expansion of NATO conventional forces which made Flexible Response feasible. Thus, by 1960, the intellectual construct for NATO's defense posture was clear, and the debates in the next three decades were more tactical than about fundamental principles.

It was in strategic systems where the most basic adjustments were made. Once Secretary McNamara had won his battles with the Congress and military, the essential features of the Triad (1300 land based missiles, increasing reliance on SLBMs, and continuous upgrades of the long-range bomber force) were in place.

The Triad and the dominance of Deterrence as a doctrine have become such basic features of U.S. strategic policy, that there is sometimes a tendency to forget what the U.S. *did not* develop and incorporate into the force structure. By accepting mutual vulnerability, the U.S. never sought to deploy weapons that could be used for a disabling "First Strike." Although the U.S. never accepted a "No First use" policy and the M-X missiles and D-5 warheads on the Trident SLBMs were sufficiently accurate to be used for attacking hardened silos, they were never authorized for procurement in sufficient numbers to be a threatening first strike force.

A second decision was to avoid building large-scale *active* ballistic missile defenses. Although one site of BMD interceptors was functioning for two years, it was seen as protection for the missile fields, not as the first step in a

---

[59]   Taylor, M., *The Uncertain Trumpet*, (New York: Harper, 1960).

population defense program. Then, once the ABM Treaty was signed, the U.S. disregarded the BMD system around Moscow and de-activated the Grand Forks site. This was certainly explicit evidence that the U.S. was relying on mutual vulnerability as its underlying strategic rationale.

Davis Bobrow has cogently summarized this process in writing, "Decisions between national security postures are bets about how best to control or influence rather distant futures. As such, they can be viewed in the framework of stimulus actions that transmit messages."[60] McNamara's principal goal was to transmit messages to the political leadership in Moscow; yet, once these basic guidelines for the U.S. force structure had been agreed upon, they sent powerful messages as well to domestic groups about where vast resources were going to be spent. The intellectual power created its own juggernaut.

There were three key groups that fought tenaciously to maintain Deterrence as the dominant U.S. doctrine and the basic rationale for the force structure. The most influential politically were the *defense contractors* who produced the offensive systems that were the backbone of Deterrence. By the mid-1960s, it was well-understood that the Triad would be a semi-permanent feature of U.S. planning, and this gave contractors for long-range bombers, ICBMs, and ballistic missile submarines a tangible stake in Deterrence. Once production lines had been established, there was a strong tendency for them to be maintained.[61] Few secretaries of defense had the political clout or will to fight the Congressional maelstrom that would ensue if a major system was phased out and not replaced by a newer variant.

Although the defense contractors were not critical in the initial decisions on Deterrence vs. Defense, once the hardware for Deterrence was purchased, there was an enormous incentive to continue. Moreover, once Deterrence systems were in place, the cost of shifting to or adding large scale BMD programs would have led to substantial criticism.

Bureaucratic support was also critical because it carried all the weight and momentum of the ways in which large organizations try to maintain themselves.[62] For the Department of Defense and its top management, these

---

[60] Bobrow, D. "Improving the Bases for Decision," in D. Bobrow, ed., *Weapons System Decisions - Political and Psychological Perspectives on Continental Defense*, (New York: Praeger, 1969), p. 4.

[61] See, J. Kurth, "Why We Buy The Weapons We Do," *Foreign Policy*, Summer 1973, No. 11, pp. 33-56.

[62] For a classic analysis of organizational behavior, see J. March and H. Simon, *Organizations*, (New York: John Wiley, 1959).

problems were compounded by the fact that the Pentagon was a monopsonist (sole buyer) for many of the systems concerned.[63] Although there have been elaborate attempts by economists to simulate the operations of a competitive market for government procurement,[64] most practitioners who have worked in the Pentagon note the strong tendency for the momentum of major systems to dominate even when the strategic environment changes considerably and force structure should be modified.[65]

Many *academic defense specialists* have also been strong lobbyists for Deterrence combined with arms control. Certainly most academics have been skeptical of the value of increased defense spending, and many had a strong conviction that moving away from Deterrence to alternative systems would be more costly and potentially destabilizing.[66] As the actual details of Soviet procurement programs in the 1970s and 1980s (before the end of the Cold War) have become widely available, more critical academic assessment have been made:

> What is most striking about the arms control experience surveyed here is what it did not do. Those who hoped arms control would bring about major reductions in existing or planned inventories or slow the introduction of new and more capable technologies have little grounds for satisfaction......Even the contributions of the ABM Treaty, arms control's chief accomplishment, are of uncertain durability. The Treaty may merely have codified the postponement of a race in defensive systems until advancing technologies made effective defenses possible.[67]

---

[63] For an overview of defense decision problems in the 1960s, see C.J. Hitch, *Decision Making for Defense*, (Berkeley, Cal.: University of California Press, 1965).

[64] For example, M. Bailey, "The Market Mechanism in the Defense Department," in R. McKean, ed., *Issues in Defense Economics*, (New York: Columbia University Press for the N.B.E.R., 1967).

[65] Kaufmann, W.W., and J. Steinbruner, *Decisions for Defense - Prospects for a New Order*, (Washington, D.C.: The Brookings Institution, 1991), p. 4.

[66] For an enthusiastic defense of Deterrence and arms control, see, A. Chayes and P. Doty, eds., *Defending Deterrence - Managing the ABM Treaty Regime Into the 21st Century*, (Washington, D.C.: Pergamon-Brassey's, 1989).

[67] Carnesale, A. and R. Haass, "Conclusions: Weighing the Evidence," in A. Carnesale and R. Haass, eds., *Superpower Arms Control - Setting the Record Straight*, (Cambridge, Mass.: Ballinger, 1987), p. 355.

These various factions pushed for different elements of the Deterrence / arms control mix, yet it is interesting to note how resilient the basic strategy was despite changing circumstances. Part of this debate reflects the inherent advantages the offense has because it can choose which targets to attack; whereas missile defense has to be designed to defend broad areas to provide successful population coverage. There were comparable rigidities at the tactical level for NATO. Once Flexible Response had been accepted in 1967, there were modest changes in burdens,[68] but very deep resistance to altering the goals during a conflict[69] or the mix of air and ground responsibilities.[70]

Hence, it is reasonable to conclude that, once Deterrence and Flexible Response were established as the dominant doctrines, a host of political, economic, and bureaucratic reasons prevented a shift toward greater uses of BMD. This was the case even when the technologies available and the geopolitical circumstances changed.

* * *

We now turn, in Chapter 3, to an examination of the current status of and prospects for theater missile defense.

---

[68] Changing the distribution of the economic burden of NATO was a notable shift. See, S. Lunn, *Burden-Sharing in NATO*, (London: Routledge and Kegan Paul, 1983).

[69] See, S. Huntington, "Conventional Deterrence and Conventional Retaliation in Europe," *International Security*, Winter 1983/84, Vol. 8, No. 3, pp. 32-56, for a discussion of how conventional force offense might have been used to deal with Warsaw Pact advantages in numbers.

[70] Canby, S. and I. Dorfer, "More Troops, Fewer Missiles," *Foreign Policy*, Winter 1983/84, No. 53, pp. 3-17.

3

Theater Missile Defense

The Issues

Theater ballistic missiles (TBMs) are usually defined as having a range of 100 to 3000 km. When most observers comment on TBMs being used in combat, they think of Soviet Scuds. This is actually quite an accurate impression because Western nations have not fired or sold TBMs that have been used in combat since the Germans launched over 2300 V-2s at the end of World War II.[1]

Since then, except for the 23 FROGs used by the Egyptians in the 1973 War, all the other combat TBMs fired have been Scuds or variants of Scuds. In the Iran-Iraq War, approximately 350 were launched; Libya used two in 1986 to retaliate for the U.S. bombing of Quaddafi's headquarters; over 2000 were used in Afghan War; and approximately 90 Scuds and modified Scuds were fired by Iraq during the 6 weeks of the Persian Gulf war in 1991.[2]

It was the success of the Patriot PAC-2 intercepts of the Scuds over Saudi Arabia that created a climate conducive to passing the Missile Defense Act of 1991.[3] Yet, the general public, and probably many in Congress, did not real-

---

[1] Soviet troops in Germany captured V-2s intact, and the Scud is an updated version of the V-2.

[2] Ballistic Missile Defense Organization, *1993 Report to the Congress on the Theater Missile Defense Initiative*, (Washington, D.C.: U.S. Department of Defense, 1993), p. 1-1.

[3] Ingersoll, B., "Patriot Missile's Success Against Iraq's Scuds Knocks Star Wars Prospects Into Higher Orbit," *Wall Street Journal*, January 29, 1991, p. A-20.

ize how different the problems are for designing missile defenses against most TBMs (that approach their targets at 3 km/second or slower) versus ICBMs that typically move more than twice as fast.

Figure 3.1 below provides a schematic diagram of the different phases of and approximate times for a TBM's flight. The Boost Phase is critical because that is the time in which the missile's rockets are firing and generating enormous heat, thus making it possible to identify its precise location with thermal imaging satellites. The second important characteristic of TBMs is that, for ranges up to 500 km, they spend much of their time inside the atmosphere (in endoatmospheric flight). A missile or separated warhead traveling *in* the atmosphere not only generates substantial frictional heat that can be tracked by thermal sensors, but also it is far easier to track by radar than a missile traveling higher and *outside* the atmosphere (in an exoatmospheric trajectory).

This means that, if a defender has a means of attacking a TBM during Boost or Mid-Course phases, the chances of identifying its location are far better than for an ICBM which travels most of its time high and in black space. Third, because the terminal velocities are slower than an ICBM, a TBM defense has more time to take repeated shots at an incoming missile or warhead.

It is these characteristics that have made specialists far more optimistic about the chances of designing and deploying effective theater BMD. Congress authorized $1.1 billion for Theater Missile Defense (TMD) efforts in Fiscal Year (FY) 1993 and the Clinton Administration, which sharply cut the rest of the defense budget, urged a 64% increase in TMD funding for FY 1994.

Nevertheless, optimism is not necessarily a sign of technical feasibility or even desirable programs. Hence, this chapter is intended to answer the following questions: How effective are current U.S. TMDs? How rapidly are TBMs spreading and what type of threat does this pose? What are the most important aspects of this for U.S. security policy? Why is there broad-based, bipartisan political support for TMD? and Will the expected improvements in TMD be able to deal with the increasing sophistication (using decoys and electronic counter measures) anticipated in future variants of attacking TBMs?

Themes of this Chapter

The basic arguments of this chapter are:

FIGURE 3.1 TBM Range and Timing Characteristics

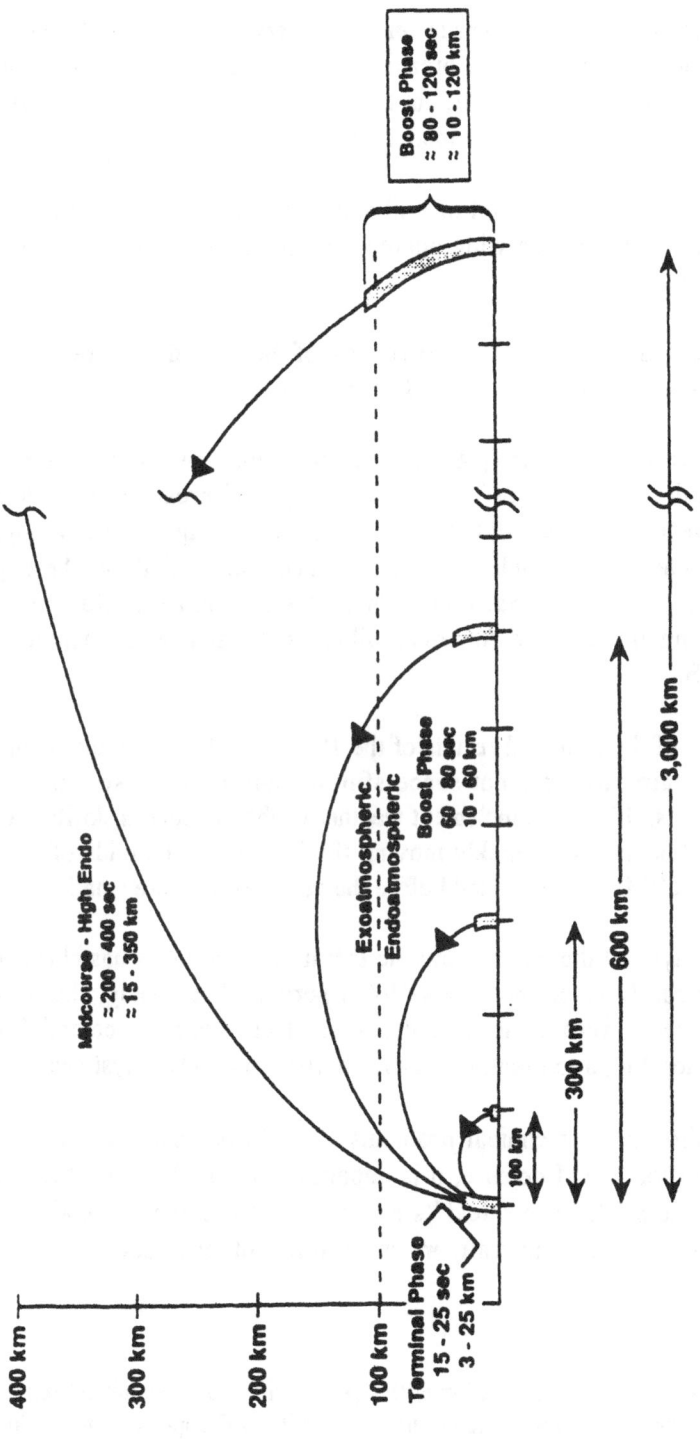

Source: 1993 Report to Congress on the Theater Missile Defense Initiative (Washington, D.C.: Ballistic Missile Defense Organization (BMDO), U.S. Department of Defense, 1993), p. 2-3.

1) There is a broad consensus among military analysts that the chances of a major global conflict breaking out have dropped dramatically since the end of the Cold War, but the perceived chances of having regional conflicts have probably increased.[4]

2) Because of the spread of ballistic missiles (pre-dominantly TBMs), there will be a growing number of nations threatened by these missiles.

3) There is no nation on the periphery of the U.S. that currently poses a direct TBM threat to the U.S. homeland.

4) However, if the U.S. plans to continue to play an activist role in world affairs, the spread of TBMs poses 3 types of threats for U.S. decision-makers: (i) a direct threat to troops, aircraft, and naval deployments abroad, (ii) an indirect threat to allied nations that may be targeted, and (iii) regional conflicts that start small could widen and eventually draw in American allies or threaten vital interests of the U.S.

5) One of the principal results of the 1991 Gulf War was the recognition that aircraft were not successful at finding and destroying mobile TBMs. If there is inclement weather or the attacker is skillful at firing TBMs and then quickly moving the launchers, the odds of knocking out TBMs on the ground before they are airborne are small.

6) The type of the ballistic missile threat the U.S. faces has changed significantly since the 1960s-1980s period. Previously, the principal concerns were Soviet ICBMs. Now, there are more potential adversaries, but, at present, they have shorter-range TBM systems.

7) Although the technical problems of TBM defense are less daunting than for ICBM defense, it is important to note that this still does not create a "defense shield" since no BMD is effective against cruise missiles, stealth aircraft, or various forms of terrorism.

---

[4] For a provocative elaboration of this point, with examples taken from Europe, see, J. Mearsheimer, "Back to the Future: Instability in Europe After the Cold War," International Security, Summer 1990, Vol. 15, No. 1, pp. 5-56.

The Scale and Significance of the TBM Threat

The immediate TBM threat is from countries that have inventories of missiles that are well-maintained and ready to use. This information is available to intelligence agencies, but it is only released, after considerable delay, to the public.[5] The striking aspect of this situation is the number of neighboring countries that are now well within range of these TBMs. Virtually every capital city in East Asia, Southeast Asia, South Asia, and the Middle East is targetable by TBMs and that was not true a decade ago. Although the present situation is troubling, in estimating the long-term TBM threat, one needs to consider not just current missile inventories but production capability and potential improvements in technology as well.

Table 3.1 below lists those nations, other than the U.S., that currently maintain TBM and ICBM production capabilities.[6] Note there are 18 countries, which makes it very unlikely that the technologies involved will be controlled; but, in addition, most have sophisticated industrial sectors and can plausibly introduce innovations that will increase the range and accuracy of their missiles.

This is the dilemma of "technological diffusion."[7] There is an incentive for countries to upgrade their own weapons, and then a host of inducements to sell the final product or the technology to others. The financial attractions of exporting missiles, components, or technology are manifold: generating foreign exchange, keeping a production line going, getting economies of scale for a new system, and encouraging a purchaser to acquire a weapon so that follow-on spare parts business is created.[8]

---

[5]   See, for example, Arms Control and Disarmament Agency, *World Military Expenditures and Arms Transfers 1988*, ACDA Publication 131, June 1989, pp. 18-19. Russia has TBMs in its central Asian area and in the Far East Military district. China is the only "developing country" that produces ICBMs. For an overview of China's missile programs, see, J. Lewis and Hua Di, "China's Ballistic Missile Programs: Technologies, Strategies, Goals," *International Security*, Fall 1992, Vol. 17, No. 2, pp. 5-40.

[6]   For our purpose here, those missiles in Table 2 with ranges of 3840 km or less will be considered the TBMs. The Chinese CSS-3 and the French M-4 are intermediate range missiles, and, under some definitions, would be defined as TBMs.

[7]   For a good discussion of the problem, see Chapt. 1, "The Challenges of Technology Diffusion," in J. Nolan, *Trappings of Power - Ballistic Missiles in the Third World*, (Washington, D.C.: Brookings Institution, 1991), pp. 1-13.

[8]   For a detailed examination of the French system of exporting arms, see, E. Kolodziej, *Making and Marketing Arms - The French Experience*, (Princeton, N.J.: Princeton University Press, 1987).

TABLE 3.1 Planned Ballistic Missile Production Ranges (in km.)

Country	120+	240+	480+	960+	1920+	3840+	7680+
Brazil	MB/EE 150 SS-150	MB/EE 300 SS-300	MB/EE 600	MB/EE 1000 SS-1000			
China	B-610	M-11	M-9		CSS-2,-5 CSS-N-3	CSS-3	CSS-4
Egypt		Scud B	Scud C	Vector			
France	Pluton	Hades			S-3,-4 M-20	M-4	M-5
India	Prithvi				Agni		
Iran	Iran-130	Scud B	Scud C				
Iraq		Scud B	Al Hussein Al Abbas	BADR 2000	Al Aabed		
Israel			Jericho 1		Jericho 2		
N. Korea		Scud B	No Dong 1	No Dong 2			
S. Korea	NHK-1	NHK-2	NHK-3,-A				
Libya				Al Fatah			
Pakistan		Hatf 2					
Taiwan	Green Bee		Sky Horse				
U.K.						A-3TK Polaris	
USSR	SS-21 M1	Scud B Scud Variant	SS-23	SS-N-5	SS-N-6	SS-N-18	SS-11,-17, -18,-19 SS-13,-24, -25SS-N-8, -20,-23

Source: *Ballistic Missile Proliferation - An Emerging Threat*, (Arlington, Va.: System Planning Corporation, 1992), p. 55.

Most nations involved in exporting weapons are choosing to do so as a result of conscious decisions by top political authorities. However, there is now a growing, legitimate concern that individuals or factions within countries will seek gain from these sales. The reports of Russian weapons specialists working overseas without authorization from Moscow and Chinese sales of M-11 missile parts that may have come from a faction of the military without full Beijing approval[9] are disturbing examples.

Nevertheless, all production and most sales of missiles are done on purpose. The widespread sale of Scuds in the Middle East was an integral part of Soviet efforts to develop and maintain client states during the 1970s and 1980s. Similarly, today many analysts see the Chinese missile connection with Pakistan as part of a long-term desire to contain India, and the reputed Chinese nuclear assistance to Syria as one step in an overall plan to court Moslem oil-exporting states.[10] In a related fashion, U.S. efforts in the 1980s to build up Iraq as a buffer against Iran had the unintended result of strengthening Saddam Hussein's military might.

A related problem is the connection between civilian space programs and military technology. Table 3.2 below lists the launch vehicles and payload characteristics for the major space systems in Russia, France, Japan, and China. In addition to these countries, India is pursuing an active space program as well. Since having the ability to loft a moderate-sized payload into polar orbit is virtually synonymous with an ICBM capability, any "collaborative" space programs will need to be watched closely. There is no evidence, to date, that space launch vehicles have been sold to nations aspiring to have ICBMs. Yet, this is not only a plausible potential development, but one that would even be internationally acceptable because it does not come under the current Missile Technology Control Regime (MTCR).

Although there may be numerous future efforts at limiting "proliferation," it seems reasonable to assume that a combination of political and economic incentives will widen even further the number of countries with operational TBM systems.

As long as TBMs have conventional explosives, they can be used to terrorize an opponent's population (as they were employed in the Iran-Iraq War), but they are not, alone, weapons of mass destruction. The greatest threat will come when a broader range of countries masters the technology necessary to put chemical, biological, and nuclear warheads on TBMs. See Appendix C

---

[9]   See, S. Holmes, "U.S. Determines China Violated Pact on Missiles," *New York Times*, August 25, 1993, p. A-1.

[10]   Safire, W., "China's Hama Rules," *New York Times*, March 5, 1992, p. A-27.

TABLE 3.2   Space Launch Vehicle Producers

Country	Operational SLVs	SLVs in Development
Brazil		VLS
China	CZ-1D, CZ-2C, CZ-2E, CZ-3, CZ-4A	CZ-3A
France	Ariane 4	Ariane 5
India	SLV-3, ASLV	PSLV, GSLV
Israel	Shavit	
Japan	M-3SII, H-I	M-5, H-II
Russia	SL-3, SL-4, SL-6, SL-8, SL-12, SL-13, SL-14, SL-16, SL-17 Energiya, SL-17 Buran, SL-11	SS-19 SLV, Sawfly (SS-N-8), Volna (SS-N-18), Shetal (SS-N-23)
Ukraine	SL-16, SL-7, SL-8, SL-14	Space Clipper (SS-24), SS-18K

Source: The Emerging Ballistic Threat to the Unites States, Report of the Proliferation
   Study Team (Washington, D.C.: U.S. Department of Defense, February 1993),
   p.3.

for data on the lethality of chemical and biological weapons.

One disturbing aspect of the Persian Gulf War of 1991 was the recognition
that a country like Iraq can make substantial strides in a nuclear weapons program while disguising its efforts and formally being a signatory to the Non-
Proliferation Treaty.[11]   This seems to be repeating itself in North Korea

---

[11]   For an overview of nuclear proliferation issues, see, L. Spector, *Nuclear
Ambitions - The Spread of Nuclear Weapons*, (Boulder, Colo.: Westview Press /
Carnegie Endowment for International Peace, 1990).

today, and there is no technical or political reason to think that it will not recur in other countries.

What is not widely understood, however, is that chemical and especially biological weapons can also be means for mass destruction. In certain regards, chemical and biological weapons require even more precision than nuclear devices in detonation and in the spread of the toxic agents. In Table 3.3 below, Steve Fetter has analyzed the comparative lethality of different types of weapons.

This may appear to be a particularly morbid sort of comparison, yet it illustrates that, under certain circumstances, biological weapons can be as devastating as nuclear ones. Obviously, TBM defense alone could not stop the use of chemical, nuclear or biological weapons. Any of these could be delivered by aircraft as well or even by skillful terrorist action, but TBMs are the easiest way for an attacker to launch such weapons.

Prior to the Persian Gulf War there was speculation about how vulnerable

TABLE 3.3   Comparing High Explosive Warheads

Type of Warhead	Without Civil Defense		With Civil Defense	
	Dead	Injured	Dead	Injured
Conventional (1 ton of high explosives)	5	13	2	6
Chemical (300 kg. of sarin)	200-3,000	200-3,000	20-300	20-300
Biological (30 kg. of anthrax spores)	20,000-80,000	0	2,000-8,000	0
Nuclear (20 kilotons)	40,000	40,000	20,000	20,000

Note: This table assumes a missile with a throw-weight of 1 ton aimed at a large city with an average population density of 30 per hectare.

Source: S. Fetter, "Ballistic Missiles and Weapons of Mass Destruction: What is the Threat? What Should be Done?" *International Security*, Summer 1991, Vol. 16, No. 1, p. 27.

TBMs would be to aircraft trying to destroy them. However, of the 2700 sorties that were devoted to finding and destroying the Iraqi Scuds, not one was a documented success.[12] This is particularly important because, after the third day of the war, the allied coalition had complete air superiority and was able to fly over most areas of Iraq at will.

Also, the U.S. had devoted exceptionally heavy satellite surveillance to the region, and few other nations could expect to have comparable reconnaissance data if they were seeking to destroy an opponent's mobile TBMs. Thus, it appears that well camouflaged, skillfully hidden, and frequently moved TBMs are survivable assets that an attacker can count on even if air superiority has been lost. This, of course, distinguishes TBMs from aircraft which are vulnerable on the ground and need at least escort or other protection to move through enemy-controlled airspace.[13] Although the ultimate effectiveness of the Iraqi Scuds was limited by the poor domestic metallurgy and manufacturing techniques used to produce the "Al-Hussein" version of the missile, that is not an inherent problem in TBMs. The survivability of the Iraqi Scuds throughout the extraordinary onslaught they faced is likely to encourage other nations to seek comparable TBMs.

The success of a weapon not only leads to the spread of that system, but to a host of imitators and further innovations as well. "For two millennia from about 500 B.C. to 1500 A.D. technological change had little impact on how wars were fought."[14] However, since then, the pace of change has accelerated so rapidly that even the wealthiest nations have trouble keeping abreast. This has been true for naval forces[15] as well as combined force operations,[16] and some analysts see technological innovation inexorably driving strategic choices.[17]

---

[12] U.S. Department of Defense, *Final Report to the Congress: Conduct of the Persian Gulf War*, (Washington, D.C.: U.S.G.P.O., April 1992), p. 159.

[13] This illustrates the sharp distinction between the Iraqi Scuds which were operated for weeks until missile reloads were exhausted versus the air force that lasted only a few days and was either destroyed or flown to Iran for safe-keeping.

[14] Weigley, R., "War and the Paradox of Technology," *International Security*, Fall 1989, Vol. 14, No. 2, p. 196.

[15] See, B. Brodie, *Sea Power in the Machine Age*, (Princeton, N.J.: Princeton University Press, 1941).

[16] Millis, W., *Arms and Men: A Study in American Military History*, (New York: Putnam and Sons, 1956).

[17] van Creveld, M., *Technology and War: From 2000 B.C. to the Present*, (New York: Free Press, 1989).

As the international competition for high technology increases and the lag time between innovation and imitation declines,[18] the spread of TBMs and the threat they pose is almost certain to grow. Although export controls on key technologies may slow the spread of TBMs and other weapons capabilities, most specialists see technological diffusion as an inevitable function of improvements in education and the speed of data transmission.[19]

Hence, it is understandable that political leaders in the industrial democracies see the growth of TBM inventories in less developed countries as a challenge and have been exploring various forms of cooperation. The Dutch, Germans, Italians, and even the Swedes have expressed an interest in improving BMD systems; and, in 1992, Boris Yeltsin even proposed a joint program to develop a Global Protection System against ballistic missiles.[20]

The Foreign Policy Context for TMD

Given the widespread agreement that the third world TMD threat has grown and that it will eventually affect both the U.S. and the other industrial nations, what are the foreign policy circumstances which make this an immediate issue?

The end of the Cold War has precipitated a vast outpouring of articles, books, and speeches about appropriate goals for the U.S. and its allies in this new environment. Given the Clinton Administration's avowed intention to focus predominantly on domestic issues, it is not surprising that there was no early statement in 1993 about U.S. foreign policy objectives. Yet, after extensive criticism about vacillation on Bosnia and uncertain objectives, the Administration did attempt to present an overview of its intentions.

On September 21, 1993, National Security Adviser, Anthony Lake, said there would be four principal goals for U.S. foreign policy: (1) "... to strengthen the core of the major market democracies, the bonds among them and their sense of common interest," (2) ... to help democracy and free markets survive in places like Russia, Eastern Europe, and other former communist

---

[18] For a quantitative assessment of this process, see, F. Scherer, *International High Technology Competition*, (Cambridge, Mass.: Harvard University Press, 1992).

[19] For a discussion of this, see, K. Bailey and R. Rudney, (eds.), *Proliferation and Export Controls*, (Lanham, Md.: University Press of America, 1993).

[20] International Study Group on Proliferation and Missile Defense, *Proliferation and Missile Defense: European-Allied and Israeli Perspectives*, (Fairfax, Va.: National Institute for Public Policy, June 1993), pp. 4-7.

lands "where we have the greatest security concerns," (3) ...to minimize "the ability of states outside the circle of democracy and open markets to threaten it," and (4) to decide where to intervene in human rights tragedies.[21]

The Clinton Administration has broadly called these goals "Enlargement," and portrays the problem of foreign policy choice as essentially providing incentives for cooperative behavior. It is obvious that these aspirations are ambitious; however, during 1993-94, the Administration's will to persist was so limited in Somalia and Bosnia and the effort devoted to Haiti was so large that many observers questioned the priorities chosen. A key problem appears to be that Clinton makes little differentiation between the desirable and the attainable. This could be very unsettling if the message received is that the U.S. is unsure about adhering to its commitments.

There is wide agreement that the post-Cold War Era is a time of broad structural shifts in international relations. Yet, there is no consensus on what the new power alignments will be,[22] and it is reasonable to assume that there will be less predictability in the behavior of small powers than there was during the period of the 1950s to 1980s. This could affect conventional military deterrence in several important ways: (a) making miscalculations more likely, and (b) making restraint or "self-deterrence" more subject to national preferences and less influenced by international norms. Hence, some nations may not be deterred by usual levels of threat and potential losses.

Two examples from the Vietnam War are worth noting. Jervis, Lebow and Stein argue:

> The U.S. policy of combining negotiations with pressure sometimes in the forms of explicit threats, more often in the form of bombing (which both inflicted pain and carried the threat to inflict more if the North continued to be recalcitrant) looks like a text book case of coercion. But the policy failed because U.S. statesmen could neither understand the North's framework of beliefs nor coordinate the complicated series of actions that were called for.[23]

---

[21] Friedman, T., "U.S. Vision of Foreign Policy Reversed," *New York Times*, September 22, 1993, p. A-13.

[22] For a detailed analysis of why the major international relations theories were incapable of anticipating the demise of the Cold War, see, J. L. Gaddis, "International Relations Theory and the End of the Cold War," *International Security*, Winter 1992/93, Vol. 17, No. 3, pp. 5-58.

[23] Jervis, R., N. Lebow, and J. Stein, *Psychology and Deterrence*, (Baltimore, Md: The Johns Hopkins University Press, 1989), p. 8.

If the U.S., with all its intelligence and analytic apparatus, made such basic miscalculations about the North Vietnamese, it is not unreasonable to think that many of the countries which have recently acquired TBMs could make comparable mistakes about their neighbors.

Kaysen, Rathjens, and McNamara point out that, despite the U.S. having a near-monopoly on "deliverable nuclear weapons," the Chinese communists continued their conquest of the Mainland, the North Koreans attacked South Korea, and the North Vietnamese continued their operations in South Vietnam. Moreover, when the French wanted help at relieving the fortress of Dien Bien Phu, they requested U.S. air strikes.

> Within the U.S. military, discussion favored the use of low-yield nuclear weapons... President Eisenhower later told his biographer that when these discussions were reported to him, he responded: 'You boys must be crazy. We can't use those awful things against Asians for a second time in less than ten years. My God.' Self-deterrence was as effective as mutual deterrence.[24]

Although most Americans today would probably agree that saving the French garrison at Dien Bien Phu was not worth using nuclear weapons, there were obviously senior officers who disagreed with Eisenhower. Can the U.S. and its main allies count on the same kind of restraint from the leadership of all the countries that have TBMs? What will happen when a growing percentage of nations acquire chemical, biological, or nuclear capabilities along with their TBMs?

These abstract concerns become specific policy problems when a nation is willing to challenge the status quo. In the discussion below, there is an attempt to identify those countries and regions where TBM issues are current policy problems.

North Korea is probably the country causing the most immediate worries among national security specialists.[25] A widely held view is that:

> The absolutist regime in the North has limited maneuvering room and must operate within very shaky military and economic structures. Although there are risks of pressing it too hard, a nuclear-

---

[24] Kaysen, C., R. McNamara, and G. Rathjens, "Nuclear Weapons After the Cold War," *Foreign Affairs*, Fall 1991, Vol. 70, No. 4, p. 100.

[25] As the situation in North Korea is changing while this ms. is being drafted, the intent here is to present the generic problems involved, not a detailed review of recent diplomatic or military moves.

armed North Korea would constitute the long-feared nightmare of the international community: an overarmed state in a desperate position; with unstable decision-makers and poor command and control.[26]

Analysts of North Korea reach these harsh conclusions because of a long history of extreme and brutal behavior by Pyongyang's leadership. Starting with the June 1950 invasion of South Korea and running through a series of grisly terrorist acts in the 1980s (including assassinating much of the S. Korean cabinet during their visit to Burma), Kim Il Sung showed a repeated willingness to flaunt international norms. This pattern is the reverse of Kaysen, McNamara, and Rathjens' concept of "self-deterrence."

In fact, during the past two years, with its skillful manipulation of both South Korea and the U.S., it might even be argued that North Korea has capitalized on the fears and uncertainty it has created in the world community.

Given the rough balance in conventional forces between North and South Korea but the far more dynamic economy in the South, most political-military analysts assume that the long-term competition on the Korean peninsula favors the South.[27] The one proviso to this optimistic forecast is that North Korea does not obtain nuclear weapons.

Decision-makers in Pyongyang appear to have fully understood the advantages to them of developing nuclear weapons, while stating otherwise. On December 13, 1991, North Korea signed a Non-Aggression Pact with the South and agreed on steps for eventual unification of the two countries. Fourteen days later, a second North / South document was signed vowing not to develop nuclear weapons and permitting extensive reciprocal inspection procedures. In January 1992, the North Koreans went even farther and agreed to a new, detailed set of inspection rules with the International Atomic Energy Agency (IAEA).

At this point, many observers thought Pyongyang's leaders had recognized the futility of trying to compete with Seoul and were just bargaining to extract concessions to deal with the dire economic circumstances in the North. This rosy perspective lost credibility during 1993 as intelligence reports showed continued high levels of activity at 7 nuclear sites in North Korea. See Figure 3.2 below for a description of these nuclear facilities.

---

[26] Bracken, P., "Nuclear Weapons and State Survival in North Korea," *Survival*, Autumn 1993, Vol. 35, No. 3, p. 137.

[27] For example, see, J. Winnefeld and J. Pollack et. al., *A New Strategy and Fewer Forces: the Pacific Dimension*, (Santa Monica, Cal.: RAND, 1992), p. 25.

FIGURE 3.2   North Korea's Nuclear Program

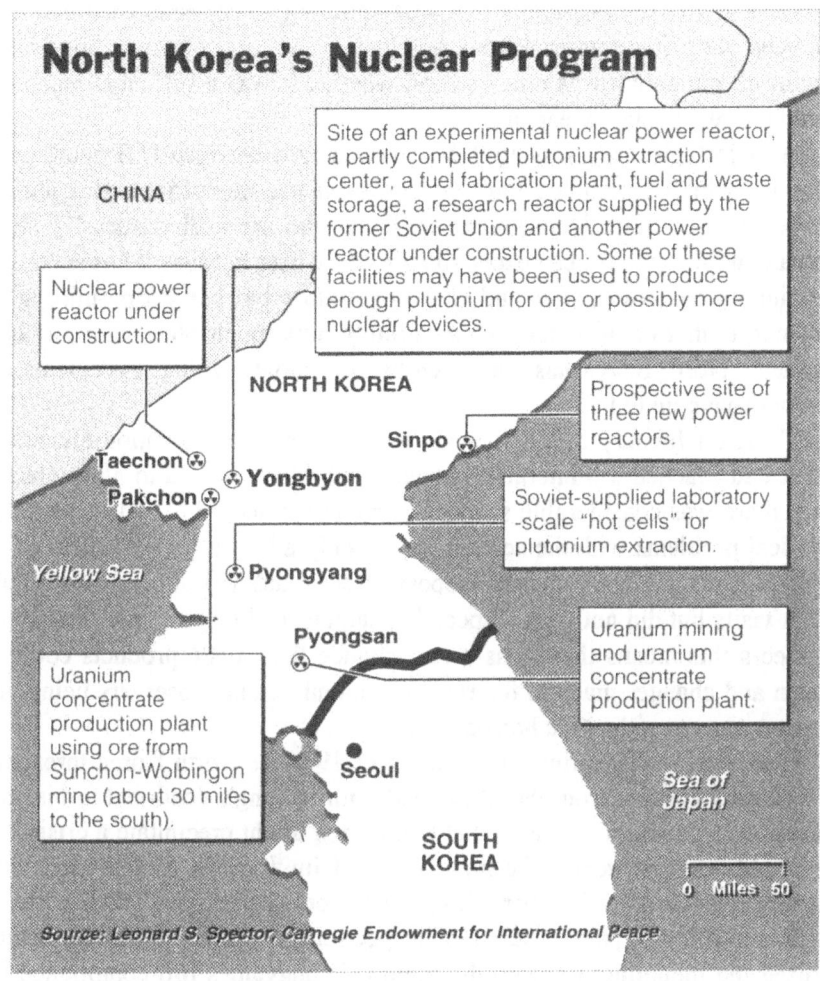

Source: New York Times, December 26, 1993, p. L-8, from L. Spector, Carnegie Endowment for International Peace. Copyright © 1993 by The New York Times Company. Reprinted by permission.

In 1989, North Korea shut down the Yongbyon reactor for approximately 100 days and IAEA inspectors (allowed in during 1992) concluded that there had been plutonium removed from the plant. What is not known, however, is how much plutonium was removed and whether it was a sufficient amount to manufacture a nuclear weapon.

A major split on assessing this issue occurred between U.S. intelligence agencies which concluded that North Korea has manufactured a nuclear weapon versus State Department analysts who are still unsure.[28] North Korean diplomacy has been exceptionally adept here because it has succeeded at stringing out both IAEA and U.S. negotiators but not yet permitting the inspections that could determine accurately how much plutonium has been diverted. North Korea has also been brazen about signing agreements and then not living up to them.

Thus, on January 5, 1994, when the Clinton Administration announced that it had reached a bilateral agreement with North Korea to allow inspections, there was considerable skepticism about the accord.[29] There were two technical problems with the accord: (a) it only allowed for inspection of the principal sites and not the waste disposal areas, and (b) it permitted regular IAEA visits but did not allow "special" unannounced inspections. To nuclear engineers this means there was a real chance that waste products could be hidden and changes made to the reactor without outside observers being able to reconstruct exactly what had occurred.

Then, during the Spring and Summer of 1994, as North Korea threatened to withdraw entirely from the IAEA and Kim Il Sung's death led to his son's accession to power, it appeared that Pyongyang might precipitate a crisis with its neighbors. However, the new leader, Kim Jong Il, took a more conciliatory line; and, by October 1994, North Korea had agreed to adhere to the IAEA and allow a phased series of inspections.[30] Although the agreement defused the mounting alarm in the region, Pyongyang's full compliance has not been confirmed, and there will always be the possibility that nuclear material has been hidden from inspectors.

----

[28]   Engelberg, S. and M. Gordon, "Intelligence Study Says North Korea Has Nuclear Bomb," *New York Times*, December 26, 1993, p. A-1.

[29]   Sanger, D., "U.S.-North Korea Atom Accord Expected to Yield Dubious Results," *New York Times*, January 9, 1994, p. L-1.

[30]   The terms of the U.S.-North Korean agreement have not been fully disclosed. Yet, apparently, Pyongyang has agreed to have the Yongbyon fuel rods examined and promised not to proceed with its nuclear program in exchange for: oil imports, external loans of $4 billion to build two light-water reactors, and improved diplomatic relations with Washington.

For purposes of this discussion, the North Korean episode illustrates several broad points: (1) programs to develop weapons of mass destruction can be effectively disguised, (2) these weapons do not have to be used or even demonstrated to extract concessions from opponents, and (3) as the range, accuracy, and sophistication of TBMs grow, the number of times when the world community faces this dilemma will also increase.

The *Middle East* is frequently suggested as the second most dangerous region of the world in terms of current TBM deployments. Iraq's success at hiding the extent of its nuclear program[31] combined with its large number of Scuds has made observers recognize that these issues are real and not just hypothetical. As Table 3.1 illustrated, there is an exceptionally wide distribution of TBM inventories in the Middle East. Also, the physical distances between many countries are small, and the complex history of national and cultural animosities makes it plausible that future conflicts will again involve TBMs as they did in the Iran-Iraq and Persian Gulf wars.

At present, there is a period of relative calm in the Middle East. The Palestinians appear to be partially on their way to statehood and Syria seems to be seriously negotiating to recognize Israel in exchange for reacquiring sovereignty over the Golan Heights.[32] Yet, the basic fissures within the region have not been healed. Saddam Hussein will certainly try to recoup his position of prominence if he can; Lebanon is still an occupied land; and most of the governments are unelected and potentially unstable.

Moreover, two very deep structural problems have not really been addressed: (a) future governance for Jerusalem,[33] and (b) the long-term links between the Moslem areas of the Former Soviet Union and the Middle East.[34] Both of these issues involve such a large number of groups with fundamentally different objectives that they seem poised to be sources of conflict even if the current agenda of high-profile topics is resolved.

Therefore, in the Middle East as a whole, the underlying problems are so pervasive and the means to inflict damage are so widely held, it would merely

---

[31] For a discussion of the problems of IAEA verification, see, K. Bailey, *Strengthening Nuclear Non-Proliferation*, (Boulder, Colo.: Westview Press, 1993), pp. 27-30.

[32] Haberman, C., "The Middle East Poker Game - After Assad's Play, the Pressure is Now on Israel," *New York Times*, January 17, 1994, p. A-6.

[33] Lustick, I., "Reinventing Jerusalem," *Foreign Policy*, Winter 1993-94, No. 93, pp. 41-59.

[34] Dannreuther, R. "Russia, Central Asia, and the Persian Gulf," *Survival*, Winter 1993-94, Vol. 35, No. 4, pp. 92-112.

be prudent to expect further use of TBMs. Also, since Israel is assumed to have nuclear weapons, there is a fundamental force imbalance which threatens all of its neighbors. Only dramatic changes in regional patterns of governance and significant arms reduction efforts would change the picture.

China, with one fifth of the world's population, is probably the next most critical area in terms of TBM development. As its economy is growing at an average of over 10% per year[35] and it shows every indication of continuing to mobilize its resources effectively, China has moved from isolation to center stage in the last decade. Some analysts now estimate that China's economy is the world's third largest, and if current growth rates continue, could be the largest by the year 2015. Few observers assume that recent levels of performance can be continued indefinitely. However, if China chooses to become a world power, it will clearly have the resources and the industrial base to do so.

The dilemma for military analysts is to ascertain what the intentions of the Chinese leadership are and to speculate on how Chinese policy might change after Deng Xiaoping passes from the scene. This process is complicated by some very contradictory signals from Beijing. Although the Chinese Army has cut its manpower levels since the decline of the Soviet threat, other actions have been notably bellicose.

The purchase of Russian SU-27s (medium range, ground attack aircraft) and negotiations with Ukraine regarding buying an aircraft carrier have been deeply troubling to China's neighbors. Also, the decision of the Chinese People's Congress on February 25, 1992 to unilaterally declare sovereignty over the Senkaku (Diaoyu Tai) and Spratley (Nang Sha) Islands was a belligerent act toward Japan and six Southeast Asian nations.

These tangible actions have also been linked with some potentially important changes in Beijing's strategy. Not since the 1600s has China had a major Navy, but, by the end of the 1980s, key military men began to talk openly about the need for a "blue water navy." In 1988, the top Chinese Admiral, Zhang Lianzhong said:

> In order to effectively protect our country from raids and attacks from the sea, we must strengthen our marine defense capacity so we can intercept and wipe out the enemy's naval forces.[36]

---

[35] See, "A Statistical Look at China - By the Numbers," *Wall Street Journal*, December 10, 1993, p. R-12, for an overview of China's economic performance data.

[36] Zhang, L., *People's Daily* (Overseas Edition), August 1, 1988, FBIS Daily Report (China), August 4, 1988, pp. 41-42.

This may sound like typical military bravado, but none of China's neighbors has a navy that is capable of attacking the mainland, and the only true naval power in the region is the U.S. That is why defense specialists found it curious that, in 1993, the Chinese military published a book titled, *Can the Chinese Army Win the Next War?* China's President, Jiang Zemin, was quoted on the cover as saying, "We must win high-tech small-scale wars under modern conditions," and the U.S. was identified as the principal military adversary in the future.[37] The book was recalled from the public, but it had an impact in the region because it identified eight war scenarios including retaking Taiwan, war on the Korean peninsula, and seizing the Spratley Islands oil fields.[38]

Growing nationalist sentiment in China is not limited to the military. Even some Western-trained foreign policy analysts are taking a harder line about projecting China's influence in world affairs.[39] These issues have been exacerbated by vacillations in U.S. positions. During 1992 many Chinese took offense at Governor Clinton's campaign rhetoric where he called for a tougher stance toward "butchers from Baghdad to Beijing." Hence, it is not surprising that Assistant Secretary of State Winston Lord was unsuccessful in May 1993 at getting concessions on his 14-point list of American concerns.[40]

Although the Clinton Administration reversed itself, seeking closer ties with China by the Fall of 1993 and dropping its threat to withhold Most Favored Nation status over Chinese human right abuses,[41] the long-run tension between Washington and Beijing seems likely to continue. Under these circumstances, China's arms exports, Beijing's unwillingness to put pressure on North Korea regarding nuclear proliferation and China's hiring of Russian weapons specialists will all heighten concerns in the West.

---

[37] Tyler, P., "Chinese Military Sees U.S. As a Foe - Disaffection in Army Seen in Book Recalled By Beijing," *New York Times*, November 16, 1993, p. A-1.

[38] One of the interpretations given for why the book was originally published was to express the resentment of China's military for the soft response given to foreign criticism after the Tiananmen Square incident and the U.S. sale of F-16s to Taiwan.

[39] See, for example, Y. Hao and G. Huan, *The Chinese View of the World*, (New York: Pantheon Books, 1989).

[40] Kristof, N., "Clinton Aide Ends Trip With No Sign of Accord," *New York Times*, May 13, 1993, p. A-10.

[41] Tyler, P., "U.S. and China Agree to Expand Defense Links," *New York Times*, November 3, 1993, p. A-13.

China, for its part, expects to be treated like a major power and has no intention of taking tutoring. Since China is both a producer and exporter of TBMs, the rest of the world will need to deal with China's missiles both as a direct threat and via its clients' capabilities. At present, with China in a cautious, non-aggressive stance, the TBM issues are a small part of a broader picture; but they would receive heightened attention if the missile exports expanded or China began to directly employ its military strength.[42] Likewise, if Beijing attempts to move from its regional power status to being a world power, this could well stimulate Japan to contain the rise of Chinese influence.[43]

Since World War II, *Japan* has very consciously foregone offensive military systems and there appears to be little public support for playing a larger military role in the world arena. Yet, there is a widespread Japanese desire for more recognition, interest in a seat on the UN Security Council, and, even among mainstream commentators, a resentment over criticism about Japan's reaction to the Persian Gulf War.[44] Thus, there is an ambivalence about being a world class economic player and continually taking the supportive and secondary role on security matters.

Nevertheless, the specter of a nuclear North Korea and a resurgent China has led the Japanese to consider establishing effective ballistic missile defenses. On September 27, 1993, Japan and the U.S. agreed to start a bilateral working group to explore cooperation on BMD.[45] As Japan already has Patriot PAC-1s, Airborne Warning and Control System (AWACS) aircraft, and is receiving deliveries of AEGIS radar and missile systems for its navy, the question becomes: what is the most efficient way to upgrade Japanese defenses?[46]

---

[42] For three scenarios of how East Asian political-military relations may develop in the 1990s, see, D. Denoon, *Real Reciprocity - Balancing U.S. Economic and Security Policies in the Pacific Basin*, (New York: Council on Foreign Relations, 1993), pp. 79-86.

[43] See, G. Segal, "The Coming Confrontation between China and Japan," *World Policy Journal*, Summer 1993, Vol. X, No. 2, pp. 27-32.

[44] For a balanced assessment of this issue, see, T. Inoguchi, "Japan's Response to the Gulf Crisis: An Analytic Overview," *Journal of Japanese Studies*, 1991, Vol. 17, No. 2, pp. 257-273.

[45] For a summary of the scope for the group and the private and public sector organizations involved, see, B. Wanner, "Washington, Tokyo Explore Defense Technology Cooperation," *JEI Report*, (Washington, D.C.: Japan Economic Institute, October 8, 1993), No. 37B, pp. 1-3.

[46] Awanohara, S., "My Shield or Yours?" *Far Eastern Economic Review*, October 14, 1993, p. 22.

The U.S. tried to get Japan to swap private sector technology in non-military fields for Patriot PAC-2s and future upgrades. However, after encountering resistance in Tokyo during his November 1993 trip, former Secretary of Defense Aspin eased the terms and apparently agreed to a straight-forward sale of improved Patriots.[47] This sets an interesting precedent because other allies will certainly want the upgraded Patriots or even more sophisticated systems that are planned for the late 1990s.

Japanese decision-makers are also being forced to think more broadly about how they wish to deal with growing challenges in their region. Ethnic Koreans living in Japan have been repatriating about $600 million per year to relatives living in the North, and this has been Pyongyang's largest single source of foreign exchange. Although the Hosokawa Government has not yet decided whether to cut off those repatriated funds, it has cracked down on firms in Japan that have been aiding the North Korean missile program.[48]

Since Japan will be co-producing the Patriots and is proceeding rapidly with its own space-launch vehicles, it would be a very easy transition for the Japanese to develop their own TBMs and ICBMs. At present, it is clearly in China's interest to avoid that, but North Korea's behavior may precipitate Japanese moves regardless of Beijing's preferences. Under those circumstances, we could see the type of TBM competition that NATO and the Warsaw Pact experienced between the late 1970s and the signing of the Intermediate Nuclear Force (INF) accords. Contemplating this possibility might either stimulate a high-tech arms race or encourage multilateral arms control negotiations in Northeast Asia.

The situation in *Eastern Europe* and the *Former Soviet Union* is less predictable than in any of the trouble spots discussed above. The Soviet Union was careful to keep tight control over the nuclear weapons and sophisticated technologies that it deployed in Eastern Europe; and most observers expect that, if instability develops in Central and Eastern Europe, it will come from ethnic and religious friction rather than arms competition per se.

States of the Former Soviet Union pose a more complex set of problems. Not only do they have the ethnic and religious friction present in Eastern Europe, but three of them (Belarus, Ukraine, and Kazakhstan) have nuclear

---

[47] Sanger, D., "U.S. Presses Japan on Missile Project - Aspin Backs Away from Requiring Tokyo to Offer Technological Secrets," *New York Times*, November 13, 1993. p. A-12.

[48] Sanger, D., "Tokyo Raids Seek to Halt Aid For North Korea on Missiles," *New York Times*, January 15, 1994, p. L-5.

weapons as well. Belarus and Kazakhstan agreed to give up their weapons quickly, but Ukraine began a long bargaining process.

In November 1993, Ukraine ratified the START Treaty and agreed to cut its nuclear arsenal of 1,656 weapons by one-third; however, both Russia and the U.S. began pushing President Kravchuk to completely eliminate these weapons as well as 175 long-range missiles. On January 10, 1994, Kravchuk agreed to do this[49] in exchange for an estimated $1 billion in payments for the nuclear material, $176 million in aid for dismantling the weapons, and $155 million in other financial assistance. Nevertheless, the agreement seems to have been a very tenuous one, as President Kravchuk refused to allow the precise terms to be made public and there was strong resistance in the Ukrainian Parliament to giving up the weapons.[50] President Kuchma, Kravchuk's successor, is less ambivalent about proceeding with removal and destruction of Ukraine's nuclear stockpile.

Yet, the uncertainties over Ukraine's plans are obviously only one piece in a bigger mosaic which will be dominated by Russia's moves[51] and the long-running debate over what types of security arrangements are most feasible for Central and Eastern Europe.[52] Although these issues do not seem as immediately volatile as North Korea's actions, if it turns out that TBMs have not been completely removed from the states of the Former Soviet Union, then stability and force balances in the region will stay in flux.

The Effectiveness of Patriot and the Role of TMD in the Current Military Environment

The initial intercept of an Iraqi Scud in Saudi Arabian skies in January 1991 gave BMD a legitimacy that it had not enjoyed before and distinguished it from the scorn heaped on President Nixon's Safeguard program and Presi-

---

[49] Apple, R., "Ukraine Gives in on Surrendering Its Nuclear Arms," *New York Times*, January 11, 1994, p. A-1.

[50] Perlez, J., "Ukraine Hesitates on Nuclear Deal - Kiev Parliament Is Reserved on Plan to Yield Weapons," *New York Times*, January 12, 1994, p. A-1.

[51] For an overview of different perspectives in the Russian foreign policy community, see, A. Arbatov, "Russia's Foreign Policy Alternatives," *International Security*, Fall 1993, Vol. 18, No. 2, pp. 5-43.

[52] The two principal models that have been suggested are: (1) a loose confederation under the Conference on Security and Cooperation in Europe (CSCE), or (2) formal military alliances like membership in NATO or some new security agreement with Russia.

dent Reagan's SDI. The deployment and functioning of the Patriots in Saudi Arabia was favorably received and warmly supported.

On the other hand, in Israel, there was more controversy over the Patriots from the start. Rather than allowing U.S. Army technicians who had been fully trained in their use to operate them, the Israel Defense Force required that its personnel be in command. There were widely reported disagreements over placement of the Patriot batteries and regarding the timing (during the descent of the incoming missile) when the interceptor should be fired.

On balance, however, the Patriots are credited with limiting panic among the Israeli public and making it possible for the Shamir Government to stay out of the broader war. This was vital for the allied coalition against Iraq because: (a) it kept the war contained in the Persian Gulf region, and (b) Israel's absence from the coalition made it politically acceptable for other Arab countries to participate.

Despite these very positive features of the Patriot performance, for the past three years there has been considerable controversy about the actual effectiveness of the interceptors. The Department of Defense did various follow-up studies and revised down the success rate estimate, but still maintains that about 52% of the Scuds were destroyed by Patriots. Both the Israel Defense Force[53] and a professor at MIT, Theodore Postol, claim that this success rate is greatly exaggerated.[54]

The essence of Postol's argument is two fold: (1) as the Scuds were descending and disintegrating many of them followed corkscrew trajectories that were hard for the Patriot radar and fire control systems to track, and (2) the incoming Scuds were moving almost twice as fast as the shrapnel from the Patriot explosives, so, unless the Patriot was right on target, the Scud sped past the defense. Postol's critique and the reply by Robert Stein, Raytheon's project manager for Patriot, have raised the level of invective on the subject.

Postol's key initial points were as follows:

1) Because of poor metallurgy, the Iraqi-modified Scud-Bs (or "Al-Husseins") frequently disintegrated on descent and the debris followed unanticipated paths.

2) The Patriot uses a proximity fuze and blows fragments toward its

---

[53] Weiner, T., "Patriot Missile's Success a Myth, Israeli Aides Say," *New York Times*, November 21,1993, p. L-13.

[54] Reuven Pedatzur, a journalist working closely with T. Postol, asserts that Israeli military sources said only one Scud was directly hit by a Patriot.

target but many of the pieces of debris were not destroyed and fell to the ground.

3) The intercept rate was very low and there may have been as few as 5 confirmed hits of the 47 Scuds fired at Israel.

4) The Gulf War experience illustrates a general problem for BMD which is that attackers can use decoys or multiple "bomblets." If the goal is "point defense" (i.e. protecting a small area), BMDs may be effective, but they have inherent weaknesses for covering larger areas.[55]

Postol also presented preliminary data on the damage caused by the Scuds and falling debris. He asserted that the 17 Scud attacks on Haifa and Tel Aviv resulted in: 1 dead, 168 wounded, 7778 apartments damaged, and 2700 people who had to be moved to different housing.

Figure 3.3 below illustrates why debris fell over such a large area. The lighter materials fell soon after being hit by the Patriot's exploding fragments, but the heavier pieces had such momentum that they proceeded onward in the direction of the original Scud trajectory. Sadly, it was the heavier pieces that did the most damage and they proceeded on into the urban areas.

The management of Raytheon thought the Postol article was so biased that they sent all subscribers to *International Security* a small brochure with a rebuttal by Robert Stein.[56] This was followed by a detailed printed letter to the Editors, published in the Summer 1992 edition of *International Security*[57]. Stein's main points in the two pieces were:

- Over 50% of the Scuds were intercepted.
- Damage to most of the apartments in Israel was superficial.
- Patriot was designed for point defense against aircraft and upgraded between August 1990 and January 1991 with software designed to deal with the Scuds. Yet, it was never meant for wide-area defense.
- No one in the U.S. Army, the Israel Defense Force, or the contractor/research community anticipated that the Iraqi Scuds would break up as they did.

---

[55]   Postol, T., "Lessons of the Gulf War Experience with Patriot," *International Security*, Winter 1991-92, Vol. 16, No. 3, pp. 119-171.

[56]   Stein, R, "Patriot ATBM Experience in the Gulf War," *Mimeo.*, (Lexington, Mass.: Raytheon Corp., Winter 1991).

[57]   Stein, R., "Correspondence: Patriot Experience in the Gulf War," *International Security*, Summer 1992, Vol. 17, No. 1, pp. 199-225.

FIGURE 3.3   Patterns of Falling Debris from Patriot Interceptors at 11 Kilometers

Altitude in Kilometers

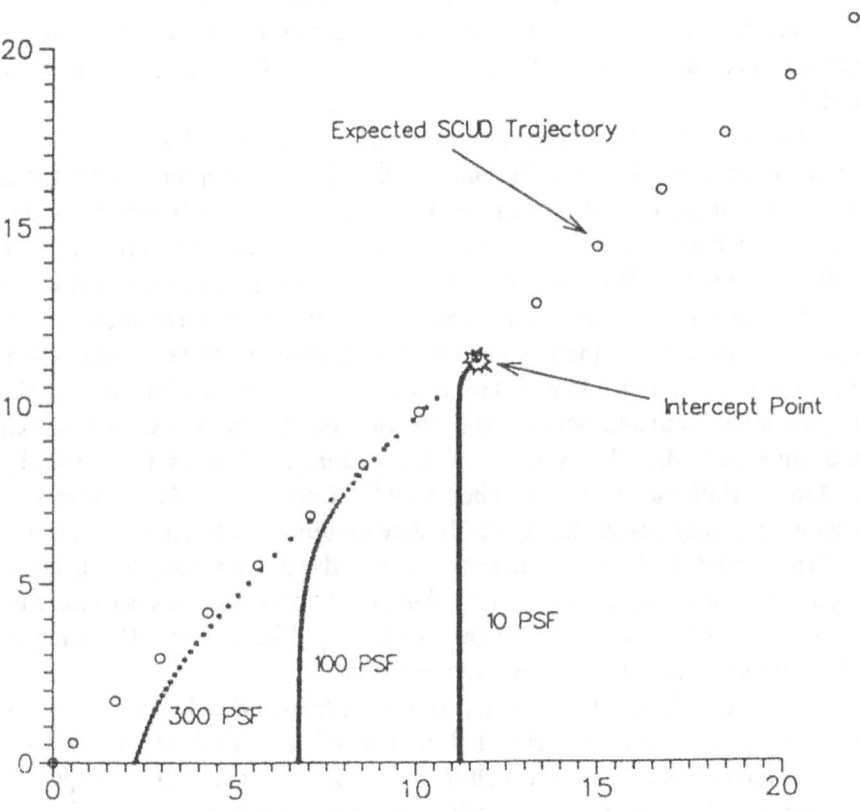

Distance in Kilometers

Source: T. Postol, "Lessons of the Gulf War Experience with Patriot," *International Security*, Winter 1991-92, vol. 16, no. 3, p. 149. Reprinted by permission.

- The Patriot intercepts in Saudi Arabia were more successful because the radars were aligned along the incoming trajectory rather than perpendicular to it as the Israelis required.

In his Summer 1992 reply (accompanying Stein's), Postol stressed: (a) the U.S. Army/Raytheon figures for actual intercepts continued to be overstated, (b) some of the Patriot misses were due to software errors, and (c) the Israeli data on Patriot performance was more credible because the Israel Defense Force did careful research on the number of firings and patterns in the debris.

The Patriot effectiveness controversy resurfaced in 1993 when Congressman John Conyers, Chairman of the House Government Operations Committee, urged the Department of Defense to have a full inquiry on the Patriot's performance in Saudi Arabia and Israel. John M. Deutsch, then Undersecretary of Defense for Acquisition and Technology, rebuffed Conyers, saying that the Army and the Patriot had performed "in an exemplary fashion" and there was no merit in holding a further investigation. Yet, staffers from the House Government Operations Committee, the Congressional Research Service, and the General Accounting Office had all been concerned when they saw the quality of analysis done on the Patriot by the Army's Ballistic Research Laboratory.[58] Because the Scuds were disintegrating as they descended, typically they presented a cluster of targets and the Army counted all cases where the warhead was disabled, destroyed or landed off-course as an "intercept." This clearly overstated the success rate for the Patriots because many of the Scuds would have been off-course or malfunctioned even if the Patriots were not fired.[59]

Since some of the data necessary to make a balanced judgment on Patriot performance in the Gulf War is still classified and other vital information may never be reconstructed, it is difficult to have a definitive view on this specific controversy. Nevertheless, some preliminary conclusions seem justified:

- All BMD systems designed to intercept warheads during their descent are, to some extent, vulnerable to decoys and the disintegrating Scud-Bs had certain characteristics like decoys.

---

[58] For a detailed review of this controversy, see, S. Hersh, "Missile Wars," *The New Yorker*, September 26, 1994, pp. 90-94.

[59] In fact, according to the photographer who took the dust jacket photos for this book, the incoming Scud pictured was not hit and it landed in a suburban area, never detonating. Under the Army's definition, this would have been an "intercept."

- This increases the urgency of designing surveillance and BMD systems that operate in the Boost Phase (before decoys are deployed or there is debris along with the warheads).
- Quicker identification of missile coordinates can be achieved through satellite surveillance as with the mini-satellites, Brilliant Eyes.
- Just because some opponents may be able to reduce BMD effectiveness (through decoys and spoofing radars) does not mean that military sites or cities would be better off having no BMD protection.

These dilemmas for missile defenses obviously need to be linked with broader questions of military strategy. This is gradually happening as analysts look at the necessary resources for upgrading missile defenses with future technology,[60] and ways to integrate TMD with other theater operations.[61] Thus, as careful attention is being given to optimal sizes for a U.S. homeland NMD,[62] it will force policy-makers to decide on the scale and sophistication that they want in TMDs. Also, both the down-sizing of U.S. military manpower and the success of high-tech weaponry in the Gulf War[63] are likely to facilitate the process of integrating plans for TMD with new surveillance and communication systems.

Current U.S. TMD Systems

Figure 3.4 below illustrates one of the most fundamental problems in ballistic missile defense. If aerial surveillance is able to accurately identify the location and direction of a missile launch, then other sensors can pick up and track the incoming missile/warheads/decoys and begin the process of discriminating between worthwhile targets and debris. This is extremely difficult in the vast, cold parts of outer space where radar and thermal sensors

---

[60] Zakheim, D. and J. Ranney, "Matching Defense Strategies to Resources: Challenges for the Clinton Administration," *International Security*, Summer 1993, Vol. 18, No. 1, pp. 51-78.

[61] Art, R., "A U.S. Military Strategy for the 1990s: Reassurance Without Dominance," *Survival*, Winter 1992/93, Vol. 34, No. 4, pp. 13-15.

[62] Durch, W., "Protecting the Homeland," Chapter in B. Blechman, et. al., *The American Military in the 21st Century*, (New York: St. Martin's Press, 1993), pp. 226-230.

[63] For a Congressional assessment of overall military performance in 1991, see, L. Aspin and W. Dickinson, *Defense for a New Era - Lessons of the Persian Gulf War*, (Washington, D.C.: Brassey's, 1992).

FIGURE 3.4 Active Defense Geometry

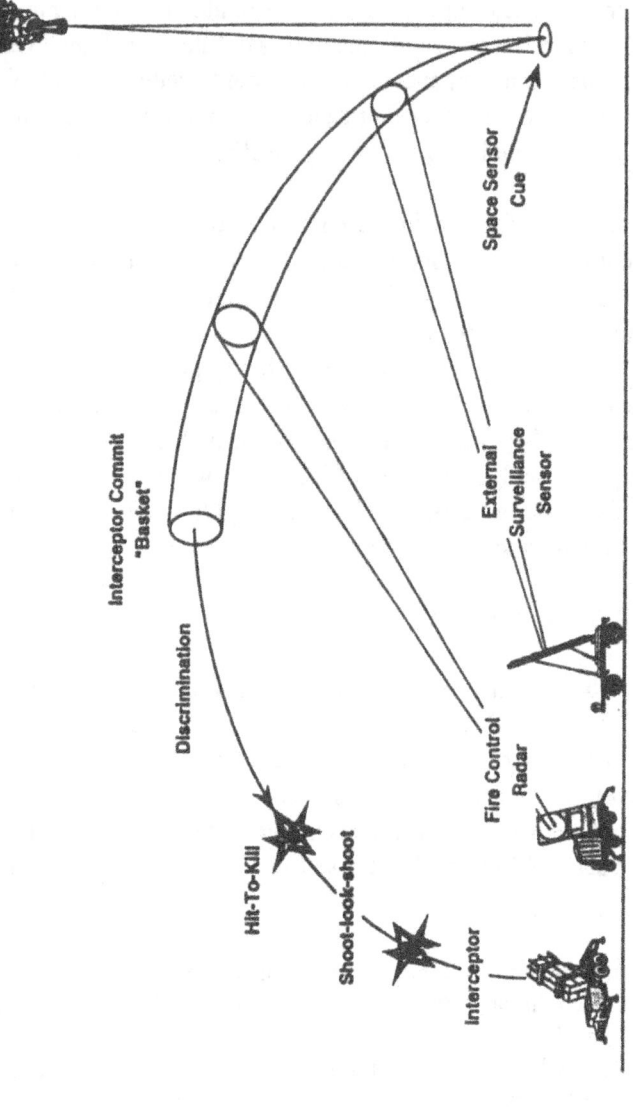

Source: 1993 Report to Congress on the Theater Missile Defense Initiative (Washington, D.C.: BMDO, U.S. Department of Defense, Washington, D.C., 1993), p. 6-9.

FIGURE 3.5 Overlap Between Theater and Strategic Systems

Derived Optimal Reentry Velocity (km/sec)

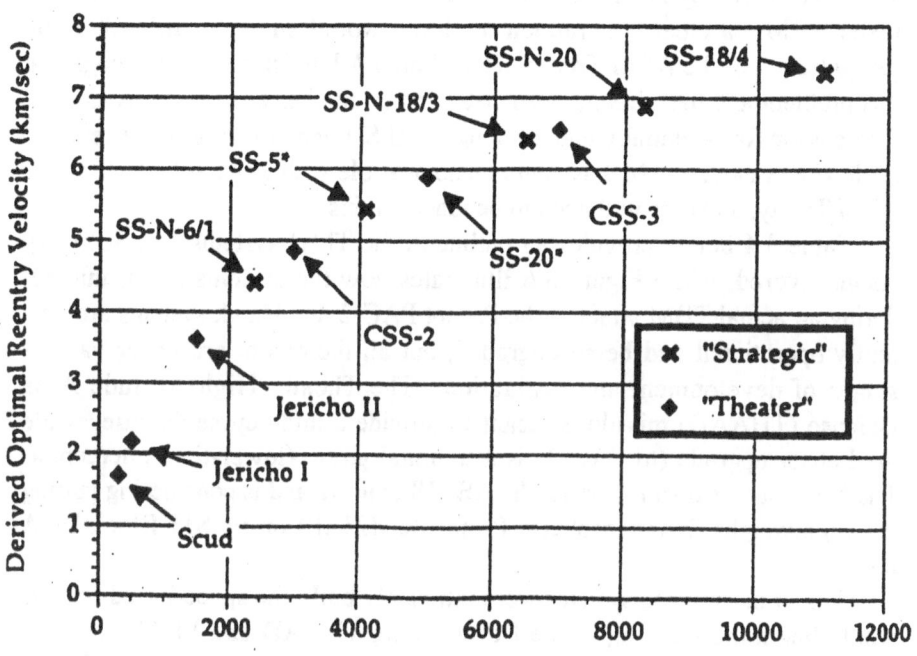

Range (km) (Min. Energy)

* Included in INF Treaty.

Source: F. Jenkins, "Implications of Defenses and the ABM Treaty for U.S. Strategic Arms Control Policy in the Nineties," *Mimeo*. (McLean, Va: Science Applications International Corporation, October 1991), p. 7. Reprinted by permission.

are less effective. Yet, for TMD, where much of the missile's flight is in the atmosphere, radar and sensors can be extremely useful if focused on the right trajectories. Hence the growing focus on Boost Phase surveillance and destruction. Additionally, boost phase interception destroys the missile before decoys have been separated from the warheads.

Figure 3.5 above illustrates a second dilemma for TMD planners: Russian and Chinese TBMs (the SS-20 and CSS-3 respectively) are now designed to enter the atmosphere at speeds faster than some ICBMs. Since effective TMD will probably need satellite surveillance[64] and many tests to deal with faster TBMs, a clear commitment to TMD would entail modification and renegotiation of the ABM Treaty. The Clinton Administration has recognized this problem and has already been criticized by the head of the Arms Control Association for potentially undermining the U.S. commitment to Deterrence.[65] As having a modern TMD system is incompatible with strict adherence to the ABM Treaty, a choice will need to be made on this.[66]

Table 3.4 below provides an outline of the TMD systems that Secretary Aspin favored, while Figure 3.6 illustrates how the systems might interact during an actual TBM attack. The Patriot PAC-2 and Hawk systems are currently operational and being upgraded, but all the others are under various stages of development and evaluation. The Theater High Altitude Area Defense (THAAD) missile is meant to provide a mid-course defense, while the Patriot upgrade (to PAC-3) is a terminal phase (lower-tier) interceptor. The Navy has its own radars on the AEGIS cruisers and is considering putting the Lightweight Exoatmospheric Projectile (LEAP) on its SM Block IV A missile.

Thus, if all of these systems were ultimately built, the space-based sensors would begin the tracking in the Boost Phase, THAAD and LEAP could be fired for mid-course defense, while upgraded Hawk and Patriot would be the terminal options for lower altitudes. Various additional systems have been suggested for actually attacking the TBM during the Boost Phase by having aircraft or drones flying with heat-seeking, air to air missiles near the battlefield.[67] As these aircraft or drones would be vulnerable themselves, it is

---

[64] In theater situations, it would be possible to use AWACS aircraft to provide mid-course surveillance, but this may not fully cover the potential launching areas.

[65] Gordon, M., "U.S. Seeking to Loosen Missile Defense Curbs," *New York Times*, December 3, 1993, p. A-14.

[66] See Chapter 6 for a full discussion of this issue.

[67] Jastrow, R., and M. Kampelman, "How to Meet the Third World Missile Threat," *Wall Street Journal*, November 19, 1993, p. A-14.

TABLE 3.4 Potential TMD Systems

Capabilities		Near Term FY 93 - 95	Midterm FY 96 - 99	Far Term FY 2000+
Lower Tier Intercept	Ground Based	• PATRIOT PAC-2 Upgrades • HAWK (USMC)	• PATRIOT PAC-3	• Corps SAM
	Sea Based	—	• AEGIS SM-2 Block IV A • SPY-1 Mod	—
Upper Tier Intercept	Ground Based	—	• THAAD (UOES) And TMD-GBR (UOES)	• THAAD (Objective) • TMD-GBR (Objective)
	Sea Based	—		• Sea Based TMD Interceptor* • SPY 1 Upgrade
Boost Phase Intercept		—	• Airborne Laser Prototype	• BPI (Objective)
Warning And Surveillance		• TPS-59 • Tactical DSP Processing	—	• Brilliant Eyes
Command, Control, Communications, Intelligence		• Launch Detection, Data Dissemination • Standardized Interfaces	• AEGIS BM / C³ Mod • Surveillance Data Netting • Communication Upgrades	• Theater Command Center Modifications • AEGIS BM / C³ Upgrades • Cooperative Engagement

Source: *1993 Report to Congress on the Theater Missile Defense Initiative* (Washington, D.C.: BMDO, U.S. Department of Defense, Washington, D.C., 1993), p. 3-9.

FIGURE 3.6 Possible Mid-Term TMD System Architecture

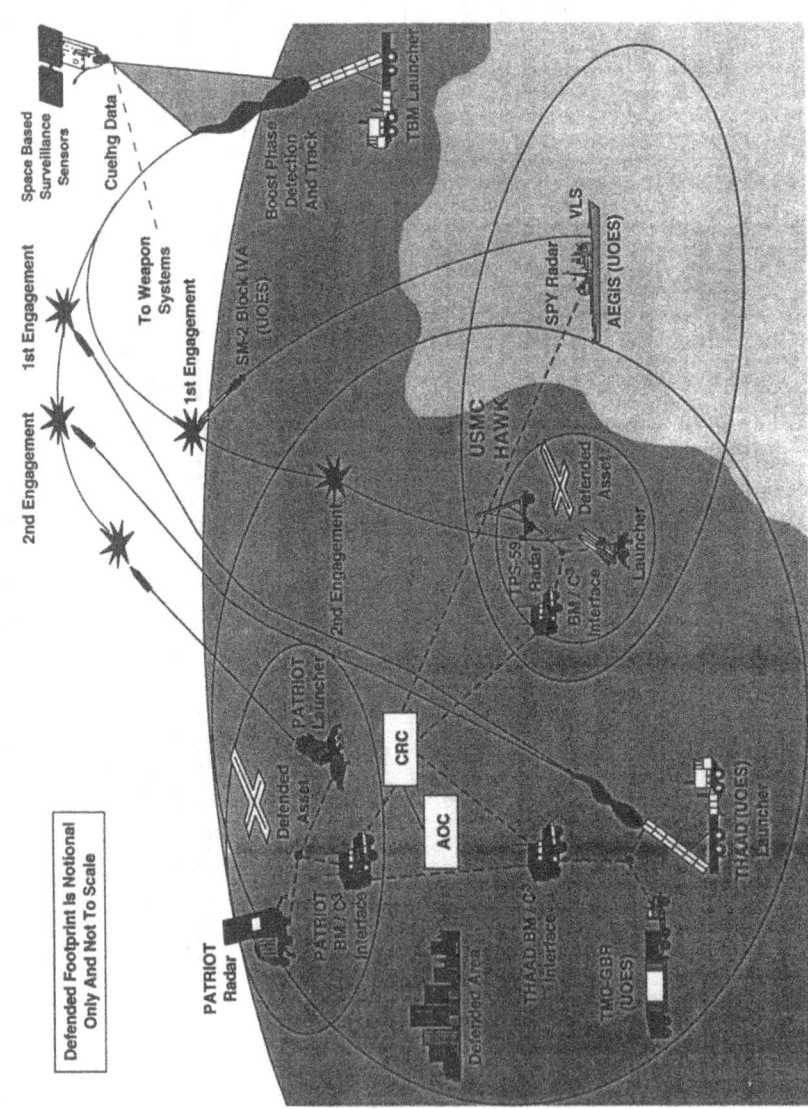

Source: 1993 Report to Congress on the Theater Missile Defense Initiative (Washington, D.C.: BMDO, U.S. Department of Defense, Washington, D.C., 1993), p. 3-10.

not clear that they would be cost-effective.

Nevertheless, there is clearly a link between having a global, space-based BMD system like GPALS (which could greatly assist in the theater operations) and long-term plans for TMD. Former Secretary Aspin specifically ruled out deploying Brilliant Pebbles (miniaturized space-based interceptors) in his Bottom-Up Review, but did acknowledge the link between TMD and strategic defense. Table 3.5 below shows various options the Clinton Administration considered. The September 1993 recommendations were to cut the Fiscal Year 1995-99 expenditures planned by the Bush Administration from $39 to $18 billion and concentrate on the "core" programs. These were: Patriot PAC-3, THAAD, AEGIS upgrades and work on battle management and communications.[68] All of these moves were reassessed by Secretary Perry in September 1994 and are likely to be re-evaluated again by the new Republican Congress in 1995.

Conclusions

1) Because the technical problems for TMD are less complex than for strategic defense, there is more confidence among the scientific and engineering community that effective BMDs can be designed and deployed.

2) In the Congress, the political support for TMD is based on an unusual amalgam of groups. Advocates of technology advancement and those generally favoring a strong defense have joined with others who have been skeptical of SDI but want to have TMD to defend U.S. troops overseas. There are also supporters of TMD who want to continue collaborative programs with NATO, Israel, and Japan in this area.

3) Academic and editorial opposition to TMD has been from several groups: some who oppose virtually all defense spending, others who want to constrain U.S. overseas operations and see TMD as a way for the U.S. to protect itself during intervention, and those who see high-performance TMD undercutting the ABM Treaty.

4) Because of the demonstrated weakness in purely ground-based TMD systems, a candid assessment of TMD requirements would acknowledge that effective TMDs pose many of the same ABM Treaty compliance problems that NMD does.

---

[68] The $18 billion figures was reached by proposing $12 on systems development and procurement, $3 billion on longer-term research, and $3 billion for general overhead.

TABLE 3.5 TMD/NMD Program Options

TMD	NMD	Technology Program		Tech Demo	Acquisition Program	
		$2 Billion Without BE	$3 Billion With BE	$7 Billion	$8 Billion	$10 Billion
Core: PAC-3	Additional TMD $14B					
THAAD	Sea-based Upper Tier $12B		12 3 (18)			12 10 (25)
AEGIS/ SM-2 Block IVA	Corps Sam $10B			10 7 (20)		
BM/C	Ascent Phase $ 9B	9 2 (15)				9 10 (22)

(NMD column shows an arrow pointing upward through $14B, $12B, $10B, $9B)

Notes: Current FY95-99 program is $39B; ◯ indicates an additional $3 billion for technology base expenditure.
Source: Data drawn from L. Aspin, "The Bottom-Up Review," *Mimeo.*, (Washington, D.C.: U.S. Dept. of Defense, September 1, 1993), p. 23.

5) As most of the industrial nations (including Russia) are now deeply concerned about the proliferation of TBMs, and there are several regions including Northeast Asia and the Middle East where the problem is also recognized, there may well be some grand bargain that could be negotiated over the future mix of offensive and defensive missile systems.

4

The Legacy of SDI

The controversy surrounding the Strategic Defense Initiative (SDI) was one of the central debates and key dividing lines in the 1980s. Most American adults had at least heard about the government's plans for 'Star Wars' and the foreign policy and defense communities were deeply split over the advisability of this venture. The legacy of SDI has not only shaped the debate over TMD and NMD in the 1990s but influenced arms control and nuclear proliferation policy as well.

When most Americans think of missile defense, they think of President Ronald Reagan. Few, in the general public, understand that the debate over SDI was "Round 2" of an intense and bitterly fought dispute that broke out in the late 1950s and reached its first fever pitch on August 6, 1969 with a 51-50 vote in the U.S. Senate[1] authorizing President Nixon to proceed with the Safeguard anti-ballistic missile (ABM) system. Although President Ford, ultimately, ordered the dismantling of the one ABM site that was constructed, opponents of missile defense formed a network that was called into action when Reagan made his March 23, 1983 speech advocating the SDI. Thus, to understand the vehemence of the SDI debate, it is useful to review some of the prior sparring on these issues.

The broad themes of this chapter are:

---

[1] The vote was on an amendment to stop the deployment of the Safeguard anti-ballistic missile system of Sprint and Spartan ground-based interceptors.

- President Reagan viewed Deterrence as dangerous and a doctrine that gave American leaders few options in a crisis situation.

- The planning for and announcement of SDI was a "top-down" effort, personally encouraged and authorized by President Reagan.[2]

- Support for strategic defense (referred to in this volume as national missile defense) was limited even during the 1984-87 period when SDI outlays reached their peak. The principal supporters were: a small number of strategists and columnists, conservative members of the House and Senate, some elements of the defense contractor community, and a limited number of appointees of the Reagan Administration.

- The opposition to SDI was widespread and included: virtually all of the arms control community, most of the media, much of the academic community, many in Congress, and substantial resistance from the Navy, Air Force, and Joint Staff.[3]

- Any radical redirection of strategy would have generated opposition, but SDI was hurt by unrealistic claims regarding its performance and frequent changes in the proposed architecture for the system.

- The SDI Organization (SDIO) was torn internally between those advocating Directed Energy Weapons (DEWs) and those who favored Kinetic Kill Vehicles (KKVs) and this led to proposals for architectures including redundant and overlapping systems.

- The Early Deployment effort (designed to get a firm commitment to proceed with NMD before the early 1990s) hurt the overall missile defense effort because the technologies were not ready and opponents used this as an opportunity to permanently limit SDI programs.

---

[2] For a detailed discussion of how the Reagan White House planned the SDI announcement and excluded the State and Defense Departments from most of the preparations, see, S. Lakoff and H. York, *A Shield in Space? Technology, Politics and the Strategic Defense Initiative*, (Berkeley, Cal.: University of California Press, 1989), pp. 7-22.

[3] For a detailed summary of the maneuvering inside the bureaucracy over SDI, see, J. Nolan, *Guardians of the Arsenal*, (New York: New Republic Books, 1989), pp. 175-229.

- By the time the SDIO had switched its support to Brilliant Pebbles (a low cost and technically efficient form of space-based kinetic energy weapon), the Cold War had abated and there was little support in either the Executive Branch or Congress for NMD.

Missile Defense Prior to 1983

The first ballistic missile attack occurred during World War II, in September 1944, when German forces fired a V-2 rocket at London. Although the Royal Air Force had learned how to foil incoming V-1s,[4] there was no adequate defense against the V-2. After the War, it was learned that the Germans had begun plans for the V-10 which was expected to have a range sufficient to hit the East Coast of the U.S.[5]

The U.S. War Department was sufficiently concerned about the prospect of ballistic missiles that it appointed one of its most prominent generals, Joseph W. Stillwell, to assess the situation. In 1946 General Stillwell concluded:

> Guided missiles, winged or non-winged, traveling at extreme altitudes and velocities in excess of supersonic speed, are inevitable. Intercontinental ranges of over 3000 miles and payloads sufficient to carry atomic explosives are to be expected...
> Guided interceptor missiles dispatched in accordance with electronically computed data obtained from radar detection stations, will be required...
> The development of defensive measures against atomic weapons should be accorded priority over all other National Defense projects.[6]

During the 1950s, the U.S. Army and Air Force competed for funding and

---

[4] The V-1s were air-breathing and, essentially, cruise missiles. They moved at low speeds and could be thwarted by barrage balloons or by aircraft that would fly alongside and force them to crash by tipping over their wings.

[5] For a discussion of this, see, D. Baucom, *The Origins of SDI 1944-83*, (Lawrence, Kan.: University of Kansas Press, 1992), p. 4, which draws on the report by General Board, "V-2 Rocket Attacks and Defense," Document 502.101-42 in the Air Force Historical Research Center, Maxwell Air Force Base, Alabama.

[6] Baucom, D., *Ibid.*, p. 6. Baucom cites as the source an appendix in, R. Jarrell and M. Cagel, *History of the Plato Anti-Missile Missile System: 1952-60*, (Redstone Arsenal, June 23, 1961).

influence in the development of ballistic missiles. The Thor and Jupiter intermediate range offensive missiles were direct competitors; and the Army developed the NIKE series of ABMs[7] while the Air Force stressed tactical aircraft for air defense. However, it was not until 1957, with the Soviet launching of Sputnik and the publication of a critical report chaired by H. Rowan Gaither, Jr. that the Defense Department began to stress its missile programs. The Gaither Report urged:

> ...the importance of providing active defense of cities or other critical areas demands the development and installation of the basic elements of a [missile defense] system at an early date. Such a system initially may have only a relatively low-level intercept capability, but would provide the framework on which to add improvements brought forth by the research and test programs.[8]

The Gaither Report was important in two regards: the visibility of its members drew attention to the subject of defense preparedness, and it gave legitimacy to the idea of deploying missile defenses and then upgrading them over time.

The Defense Department eventually clarified bureaucratic responsibilities, leaving medium and long-range offensive missile development to the Air Force[9] and defensive systems to the Army. Although the Army's initial ABM system, ZEUS, proved unable to differentiate incoming warheads from decoys, the development of phased array radars gave the next generation, NIKE-X, significantly more discriminant ability and greater accuracy.

In 1958 the Advanced Research Projects Agency (ARPA) also began funding an interesting mix of projects, such as work on boost intercept techniques, lasers and particle beams, and satellite tracking of incoming warheads.[10] Each of these concepts was used, with up-dated technology, twenty-five years later in the SDI program.

---

[7] The Army's ABM systems grew out of their NIKE anti-aircraft missiles. The chronological order for ABM systems then became: NIKE-ZEUS, NIKE-X, SENTINEL, and SAFEGUARD.

[8] U.S. Executive Office of the President, Security Resources Panel of the Science Advisory Committee, *Deterrence and Survival in the Nuclear Age: Report to the President*, (Washington, D.C.: November 7, 1957), p. 19.

[9] The Navy, of course, was developing the Polaris (and follow-on systems) but they were specifically designed for submarines, not general-purpose offensive missiles.

[10] The directed energy work was under Guidelines Identification Program for Antimissile Research (GLIPAR), the boost phase intercept research through Ballistic

Nevertheless, early BMD efforts faced a fundamental barrier. If offensive missiles deployed decoys once they were out of the atmosphere, the decoys could travel in a cloud shrouding the warheads. Most decoys would be light-weight and would, therefore, separate from the warheads only after hitting the atmosphere as they neared their targets. This gave ground-based interceptors only a brief minute or so to attack the warheads before they landed.

In a highly classified 1958 panel report, chaired by William E. Bradley,[11] the mathematics of missile defense was presented. It was estimated that the USSR could overwhelm even a multi-site ground-based ABM system if the attacks were from different directions and used large numbers of decoys. The report concluded that the U.S. needed either to: (a) get far more accurate discrimination techniques, or (b) be able to stop opponents' missiles before decoys were deployed. Those basic issues still remain with us today and formed much of the basis for debate about the advisability of SDI.[12] In 1963, Secretary of Defense McNamara decided against further funding for ZEUS, but supported the development of the NIKE-X.

The leaders of the Soviet Union did not share the U.S. ambivalence about ABM systems.[13] In fact, few Americans realize Moscow has had an opera-tional ABM system for 25 years, and that many in the Kremlin have con-sistently favored missile defense over pure Deterrence. The following state-ment by Soviet Premier Kosygin, in 1967 before the Glassboro Summit, illustrates this skepticism about Deterrence:

> Which weapons should be regarded as a tension factor - offensive or defensive weapons? I think that a defensive system, which prevents attack is not a cause of the arms race but represents a factor prevent-

---

(Continued)
Missile Boost Intercept (BAMBI), and the tracking effort through Interception By Satellite Tracking (INSATRAC).

[11] Bradley was a member of President Eisenhower's Science Advisory Commit-tee, and the panel was called the Reentry Body Identification Group (RBIG).

[12] One of the attractions of space-based ABM systems is that they can fly within a "line of sight" view of missiles being launched and thus can destroy a missile dur-ing the Boost Phase, before decoys are deployed. One key disadvantage, however, of low-altitude satellites is that the curvature of the Earth blocks much of their sight or target range. Thus, for example, if a satellite can only see 1/10 of the earth's surface on a particular path, then nine other satellites will need to be deployed to keep that particular path covered at all times. This "absentee ratio" problem is a key cost factor in designing space-based ABMs.

[13] For an overview of Soviet investments in missile defense, see, W. Van Cleave, *Fortress USSR*, (Stanford, Cal.: Hoover Institution Press, 1986).

ing the death of people. Some persons reason thus: Which is cheaper, to have offensive weapons that destroy cities and entire states or to have defensive weapons that can prevent this destruction? At present the theory is current in some places that one should develop whichever system is cheaper. Such "theoreticians" argue also about how much it costs to kill a person - $500,000 or $100,000? An antimissile system may cost more than an offensive one, but it is intended not for killing people but for saving human lives. [14]

In the week prior to the June 23, 1967 Glassboro Summit, the Chinese government announced that it had successfully tested a hydrogen bomb. This put pressure on Premier Kosygin and President Lyndon Johnson to discuss arms control, missile defense, and different types of deterrence. Yet, it was clear to all involved that the USSR was unwilling to give up its ABM. Secretary McNamara was very explicit in telling Kosygin that, if the Soviet Union chose to deploy more ABMs, the U.S. would simply add offensive missiles. This appeared to infuriate Kosygin and may have even delayed the start of U.S.-Soviet arms control talks. [15]

This led to a curious episode in American strategic thinking. President Johnson authorized Secretary McNamara to give a speech explaining why the U.S. would proceed to deploy a light ABM. Johnson's intent was to have sufficient missile defense to deal with a possible Chinese attack, [16] and to lay the basis for a comprehensive system that could, ultimately, defend against the Soviet Union as well. [17]

Secretary McNamara's September 18, 1967 speech put the issue in a totally different light than President Johnson expected. Although at the end of the speech McNamara did make the case for a light ABM, most of the presentation dealt with the high cost and futility of missile defense. [18] Despite

---

[14] Aleksei N. Kosygin as cited in B. Adams, *Ballistic Missile Defense*, (New York: American Elsevier Publishing Co., 1971), p. 154.

[15] Newhouse, J., *Cold Dawn - The Story of SALT*, (New York: Holt, Rinehart and Winston, 1973), p. 205.

[16] The Cultural Revolution had started the year before, and Mao Tse-Tung had been widely quoted as saying the U.S. was a "paper tiger," so there was broad support for having some protection against the Chinese.

[17] Halperin, M., "The Decision to Deploy the ABM: Bureaucratic and Domestic Politics in the Johnson Administration," *World Politics*, October 1972, Vol. 25, p. 87.

[18] For a text of the speech, see the *New York Times*, September 19, 1967, pp. 18-19.

criticism, President Johnson chose to proceed with the SENTINEL system[19] anyway, but his most prominent defense adviser had gone public with the case against the ABM.

The Johnson Administration then followed a two-track policy of attempting to proceed with arms control and, at the same time, beginning deployment of an ABM. The U.S.-Soviet arms control talks planned for September 1968 were postponed because of the Soviet invasion of Czechoslovakia. Yet, the negotiations finally got underway in 1969, under President Nixon, as the SALT I talks.

Before turning to the early opposition to missile defense, it is worth noting that *passive defense* (hardening or hiding offensive systems) has been relatively uncontroversial in the U.S. The development of submarine launched ballistic missiles (SLBMs) received broad support inside and outside the government.[20] Similarly, hardening of ICBM silos, first proposed by a University of Michigan group as a substitute for ABMs,[21] has been seen as stabilizing and supported by many groups that typically oppose defense expenditures.

Richard Perle, former Assistant Secretary of Defense, comments:

> We long ago realized that undefended offensive weapons could not constitute an effective deterrent. So we have, for many years and in a great many ways, sought to defend our strategic retaliatory forces. We pour tons of concrete around our missile silos and fit them with shock absorption devices so they can survive a nearby nuclear detonation. We put missiles on submarines and hide them under the seas so they cannot be found and destroyed. We place aircraft on alert so they can fly away on warning and thus escape attack...
>
> All of these passive defenses are intended to reinforce the strategy of deterrence. Indeed, without them our triad of strategic retaliatory forces would be more of an invitation to attack than a deterrent against it.[22]

---

[19] SENTINEL consisted of both endo- and exo-atmospheric ground-based interceptors. President Nixon renamed it SAFEGUARD and proposed deployment in several stages, but it was essentially the same ABM hardware.

[20] For an overview of the first SLBM program, see, H. Sapolsky, *The Polaris System Development: Bureaucratic and Programmatic Success in Government*, (Cambridge, Mass.: Harvard University Press, 1972).

[21] *Congressional Record*, 91st Congress, 1st session, July 9, 1969, 115: 18910.

[22] Statement by Richard Perle before the Committee on Armed Services, U.S. House of Representatives, April 16, 1991.

Thus, in the U.S., we have a strange situation where the commitment to ensuring the success of offensive weapons goes uncontested, whereas efforts at missile defense are considered heretical.

Despite the assumption in the Stillwell and Gaither reports that BMD would be a natural and desirable step, strong opposition to missile defense had a 25 year history before President Reagan's March 1983 speech. The earliest opposition came on purely technical grounds: Would BMDs actually work? and How effective would they be?

As mentioned above, the Bradley Report demonstrated that decoys could nullify the effectiveness of the radars and ground-based interceptors available in the late 1950s. Although phased array radars could improve discrimination, they by no means solved the problem. Since early BMDs relied on nuclear explosions, BMD development necessitated atmospheric testing to evaluate the properties of the shock, heat and electromagnetic pulse that was generated. This created great controversy in the scientific community; and, in 1964, when the former Science Adviser to President Kennedy and the former Director of ARPA both took a strong position opposing BMD,[23] this step helped mold the core of the group that tenaciously opposed the SAFEGUARD system in 1969.

One of the most vitriolic strategic debates concerned whether it was cheaper to add offensive or defensive capability. By January of 1967, McNamara's staff had completed studies which showed that the U.S. was capable of building relatively effective NMD as long as the Soviet Union was constrained to moderate increases in its offensive weapons budget.[24] It was estimated that a deployed SENTINEL system would save approximately 100 million lives if the USSR struck first in a nuclear attack on U.S. cities and military installations; but that the effectiveness of U.S. NMD would go down sharply if the Soviets added MIRVs and sophisticated decoys.

Opponents of BMD argued that, because SENTINEL was not 100% effective and its performance would decline as the USSR modernized its missile force, it was better to emphasize offensive systems rather than defense.[25]

---

[23]   H. York and J. Wiesner, "National Security and the Nuclear Test Ban," *Scientific American*, October 1964, pp. 27-35.

[24]   See, "Statement of Secretary of Defense Robert S. McNamara before a Joint Session of the Senate Armed Services Committee and the Senate Subcommittee on Department of Defense Appropriations on the Fiscal year 1968-72 Defense Program and the 1968 Defense Budget," January 23, 1967.

[25]   For a popularized version of this argument, see, J. Wiesner, "The Case Against an Antiballistic Missile System," *Look*, November 28, 1967.

Supporters of missile defense felt this ignored the tangible benefits that would come from saving most if not all of a population.[26] Some recent analysts have now claimed that the opponents misrepresented the situation in the late 1960s by leaving the impression that cost effectiveness ratios always favored missile offense over defense.[27] For our purposes, however, the key point is that the opposition to BMD was highly visible, well-organized, and had sufficient credibility to challenge SAFEGUARD in 1969 and SDI in the 1980s. It is with this background that we now turn to President Reagan's efforts to develop a comprehensive missile defense for the U.S. population. See Appendix A for an overview of the U.S.-Soviet arms control agreements between 1972 and 1979, and Appendix B for an explanation of changes in Soviet BMD doctrine and programs.

SDI in 1983 and the Immediate Reactions

When President Reagan launched the Strategic Defense Initiative, he had three basic motivations: (a) to give himself and future presidents more options than Deterrence provides in a nuclear crisis, (b) to establish a morally preferable strategy of "defending rather than avenging lives," and (c) to compete with the Soviet Union in an arena where U.S. technology was, on balance, superior.

Some journalists have claimed that President Reagan was duped into supporting SDI by false claims from Edward Teller about the likely effectiveness of the "X-Ray Laser."[28] Although Reagan may have been intrigued by Teller's hopes, the SDIO never based its architecture on the X-Ray Laser; and the President's objections to Deterrence were much more personal than technical. He actually believed missile defenses were desirable. Having had no prior experience in foreign affairs, he had been shocked to learn in 1979, during a visit to command facilities in Colorado, that the U.S. had no missile defenses. Reagan saw this as irresponsible policy; and, on numerous occasions, expressed to his close confidantes that he wanted to change this.[29]

---

[26] See, for example, F. Dyson, "A Case For Missile Defense," *Bulletin of the Atomic Scientists*, April 1969, Vol. 25, p. 32.

[27] Goldfischer, D., *The Best Defense*, (Ithaca, N.Y.: Cornell University Press, 1993), p. 170.

[28] For an exhaustive presentation of this argument, see, W. Broad, *Teller's War - The Top Secret Story Behind the Star Wars Deception*, (New York: Simon & Schuster, 1992), pp. 94-137.

[29] For a full description of Reagan's visit to NORAD and the deep impression it made upon him, see, M. Anderson, *Revolution - The Reagan Legacy*, (Stanford,

Moreover, between the SAFEGUARD debate of 1969 and 1983 there had been a number of technological advances that promised to make several types of BMD more effective. Computing speed had been dramatically increased, there had been stunning improvements in signal processing and imaging, and vast strides were made in the miniaturization of electronic circuitry. The combination of these advances made it possible to have weapons that were smaller and more accurate. Although these same technical changes could help in creating more maneuverable offensive weapons, on balance, the changes helped the defense.

Also, it was clear that the U.S. had a commanding lead over the Soviet Union in the areas that were most critical for developing the next generation of anti-missile weapons. The U.S. was far ahead in precision-guided and terminally-guided conventional systems;[30] and, though the USSR was seen to be leading in space propulsion and certain directed energy weapons, American technology was overall superior. For an administration committed to out-competing the Soviet Union, SDI was one of many routes to its foreign policy goals.

Shortly after President Reagan's March 23, 1983 speech, the Fletcher Commission was established to set preliminary goals and scientific requirements for an R & D program to lay the basis for an integrated NMD system. The Fletcher Report, completed in February 1984, divided the functions of NMD into a number of distinct categories: surveillance and acquisition, tracking, discrimination, kill assessment, battle management, fire control, target kill, and power and materials. The Commission suggested a potential architecture for the system and concluded that NMD had sufficient promise to warrant a major research and development program.[31]

Two related government-sponsored groups looked at other aspects of the problem. Fred Hoffman chaired a study which evaluated how BMD systems would affect strategy for the U.S., allies, and Soviet Union;[32] and Frank Miller chaired an interdepartmental assessment of how U.S. NMD would affect

---

(Continued)

Cal.: Hoover Institution Press, 1990), p. 82.

[30] Examples of such systems that were fully operational in 1983 were: Stinger and the Multiple Launch Rocket System with Terminally-Guided Warheads.

[31] The Commission was named for James Fletcher, former Director of the National Aeronautics and Space Administration. See, J. Fletcher, et. al., *The Strategic Defense Initiative - Defense Technologies Study*, (Washington, D.C.: U.S. Department of Defense, April 1984).

[32] Hoffman, F., et. al., *Ballistic Missile Defense and U.S. National Security*, (Washington, D.C.: U.S. Department of Defense, 1983).

arms control agreements and negotiations.[33] Subsequently, three governmental reports were issued dealing with aspects of Soviet missile defense and space programs.[34]

The early goals of the Strategic Defense Initiative Organization (SDIO) reflected both Reagan's vision and the Fletcher Commission's conception that NMD should be comprehensive, protecting both population centers and military facilities. There was also full acceptance of the need to evaluate space-based weapons which, if deployed, would have been a clear violation of the ABM Treaty and the UN Resolution on Non-Militarization of Space. This brought a hailstorm of criticism from both the scientific and arms control communities. The resistance to SDI was concentrated in three main areas: its technical feasibility, the likely response from the USSR, and the cost.

The technical feasibility was challenged on several accounts. Some computer software specialists claimed that the system would be so complex that writing the instructions code would lag far behind the weapons development.[35] Former Secretary of Defense Harold Brown asserted that "a boost phase or post-boost phase intercept is not a realistic prospect in the face of likely offensive countermeasures and the vulnerability of those tiers to defense suppression."[36] Additionally, there was intense skepticism about when the U.S. would really have operational Directed Energy Weapons (DEWs), and this led to a multi-year study by the American Physical Society that cast doubt on the near-term prospects for DEWs.[37] Although each of these technical criticisms was rebutted by various advocates,[38] the attacks forced some

---

[33] The Miller study grew out of an inter-agency staff working group and it is still classified, as are the details of the Hoffman Report.

[34] Defense Intelligence Agency, *Soviet Military Space Doctrine*, (Washington, D.C.: D.I.A. # DDB-1400-16-84, August 1, 1984,), C. Weinberger and G. Shultz, *Soviet Strategic Defense Programs*, (Washington, D.C.: U.S. Departments of State and Defense, October 1985), and U.S. Department of Defense, *The Soviet Space Challenge*, (Washington, D.C.: November 1987).

[35] Winograd, T., "Strategic Computing Research and the Universities," *Mimeo.*, (Stanford, Cal.: Stanford University, November 10, 1986).

[36] Brown, H., "Is SDI Technically Feasible?" *Foreign Affairs - America and the World 1985*, Vol. 64, No. 3, p. 447.

[37] For a summary of the report, see, "Report to the APS of the Study Group on Science and Technology of Directed Energy Weapons," *Physics Today*, May 1987, pp. S-3 to S-16.

[38] See, for example, the critique of the APS Report by L. Wood and G. Canavan, "Statement on the American Physical Society Report," presented to the House Republican Research Committee, *Mimeo.*, May 19, 1987, pp. 1-10.

reconsideration of the underlying conception in the Fletcher Commission's proposed "notional architecture." The Fletcher Report consisted of seven volumes. Volume VI discussed systems concepts and recommended a multi-tiered approach to missile defense, but it did not formally recommend a specific architecture.

Some of the most heated debate, however, regarded the likely Soviet response. Several of the key opponents to SAFEGUARD joined together again and claimed that, if the U.S. deployed a space-based NMD, the USSR would: convert its missile force to fast-burn boosters, possibly spin them to minimize the damage from lasers, and / or attack the lasers themselves.[39] Other scientists argued that "the program is too large, too political, and raises false hopes" about security.[40]

This was similar to the reaction from the arms control community which was apoplectic. Many said that SDI would undercut strategic stability,[41] that it might lead to a doubling of the Soviet ICBM force,[42] and that Moscow would escalate the quality of its offensive systems, possibly by increasing the number of its MIRVs and their accuracy.[43]

Opponents also claimed that SDI would be unaffordable. Former Secretary of Defense James Schlesinger pronounced that it would cost a trillion dollars, and the Union of Concerned Scientists estimated that there would need to be 2400 space-based satellites to do an effective job of boost phase intercept.[44] When more careful opponents estimated the costs, they looked at four potential SDI systems and concluded that they might range in expense from $160 to

---

[39] Bethe, H., R. Garwin, K. Gottfried, and H. Kendall, "Space Based Ballistic Missile Defense," *Scientific American*, October 1984, pp. 39-49.

[40] Panofsky, W., "The Strategic Defense Initiative: Perception vs. Reality," *Physics Today*, June 1985, pp. 34-45.

[41] See, for example, S. Wells and R. Litwak, (eds.), *Strategic Defenses and Soviet-American Relations*, (Cambridge, Mass.: Ballinger, 1987) and S. Miller and S. Van Evera, (eds.), *The Star Wars Controversy*, (Princeton, N.J.: Princeton University Press, 1986).

[42] Carter, A., and D. Schwartz, (eds.) *Ballistic Missile Defense*, (Washington, D.C.: Brookings Institution, 1984), p. 101.

[43] Drell, S., P. Farley, and D. Holloway, *The Reagan Strategic Defense Initiative: A Technical, Political, and Arms Control Assessment*, (Stanford, Cal.: Stanford Center for International Security and Arms Control, 1984).

[44] See, J. Tirman, (ed.) *The Fallacy of Star Wars*, (New York: Vintage Press, 1984). Later, Richard Garwin, one of the UCS leaders, admitted that they had miscalculated and that the actual number of lasers needed was closer to 100; see, *Physics Today*, March 1986, p. 148.

$770 billion.[45] Interestingly, some of the most prominent critics said that SDI "offers no prospect for a leak-proof defense"[46] and neglected to mention that SDI could have imposed considerable costs on the Soviet Union (if the USSR chose to upgrade or modify its offensive systems). Nevertheless, the critics shook public confidence in SDI and it forced a series of basic changes in the architecture being considered. The SDIO did not actually have an architecture approved by the Defense Acquisition Board until the Fall of 1987, but the critics did not wait in their attempt to rally public opinion.

Changes in the Architecture for SDI

One of the fundamental problems with the SDI was the frequent scuttling of prior plans and bold announcements about new directions the program would take. Not all of this was due to the management of the SDIO.[47] When Gen. Abrahamson took over the SDIO, he was presented with the preliminary findings of the Fletcher Commission and he was immediately pressed to justify or reject the "notional architecture" that its 60 members had proposed. Moreover, most of the critics cited above saw no reason to wait until the SDIO proposed a specific plan to begin their attack on NMD. Thus, it is worthwhile to review how dramatically the basic plans for SDI were changed between 1984 and 1991.[48] Throughout the period, the SDI assumed the U.S. would be deploying space-based systems but there was a steady move away from plans to use DEWs to support for various types of Kinetic Kill Vehicles (KKVs).

The most basic goal of the Fletcher Commission was to evaluate whether comprehensive, population defense was sufficiently feasible to justify a full

---

[45] Blechman, B. and V. Utgoff, "The Macroeconomics of Strategic Defenses," *International Security*, Winter 1986-87, Vol. 11, No. 3, p. 53.

[46] See, for example, McG. Bundy, G. Kennan, R. McNamara, and G. Smith, "The President's Choice: Star Wars or Arms Control," *Foreign Affairs*, Winter 1984/85, Vol. 63, No. 2, p. 265.

[47] While the Fletcher Commission was proceeding, the SDIO was formed by a National Security Decision Directive in January 1984. In March 1984, Lt. Gen. James Abrahamson was named to be the Director, and in April 1984 the SDIO was formally chartered by Secretary Weinberger.

[48] As will be covered in Chapter 5, when President Bush endorsed the Global Protection Against Limited Strikes (GPALS) concept in January 1991, he was altering the basic goals of SDI which had been to provide a comprehensive missile defense against a large-scale attack.

scale R & D program. Once it was decided that there was at least sufficient promise of comprehensive missile defense to warrant a major R & D program, then the commission focused on alternative ways in which this might be accomplished and principles that should guide the research.

The commission made many suggestions, but four of them were critical for the future development of SDI: (a) favoring a "multi-tiered" defense, (b) ensuring that the NMD components themselves were survivable against attack, (c) designing the capabilities to deal with the hypothesized future Soviet missile force, not just the current one, and (d) making sure that human commanders could control the system at all times.

Figures 4.1 and 4.2 below are taken directly from the Fletcher Commission Report and they give an idea of the early conceptions for a multi-tiered defense. Not only were different types of space-based weapons anticipated, but defensive measures considered necessary to stop anti-satellite (ASAT) attacks from the opponent.[49]

Although many in the Reagan Administration were satisfied with the Fletcher Commission conclusions, some of the strongest advocates for missile defense thought the report, ultimately, constrained deployment efforts.[50] By arguing that deployment should not proceed unless the system could deal with the future or 'responsive threat,' there would always be the uncertainty about whether the system under consideration met future needs. This would give opponents an easy way to block a system if they could speculate on counter-measures that were conceivably usable in the future.

Additionally, the Fletcher Commission wanted to reassure the public that an NMD system would not function without human activation and control. This seemed like a cautious criterion, but, in the initial design stages, it vastly complicated the "battle management" function because planners felt it was necessary to have "manual override" on the performance of each component. This meant a slower, less resilient system because all the parts needed to communicate with and be controlled from ground stations.

Therefore, Gen. Abrahamson had to build an organization and decide what initial system he favored, while simultaneously being criticized by a broad range of opponents. 1984 and 1985 were thus a time when SDIO was attempting to establish itself and decide on what architecture to pursue. The

---

[49]   Thus, the Fletcher "notional architecture" included: space-based sensors, DEWs, KKVs, satellite defenders, and ground-based radars and interceptors.

[50]   For an elaboration of this view, see, A. Codevilla, "Who Killed SDI? We could have had a working anti-missile defense 30 years ago. We still don't. Why?" *National Review*, May 10, 1993, pp. 40-43.

FIGURE 4.1   Fletcher Report: Phases of Ballistic Missile Trajectories

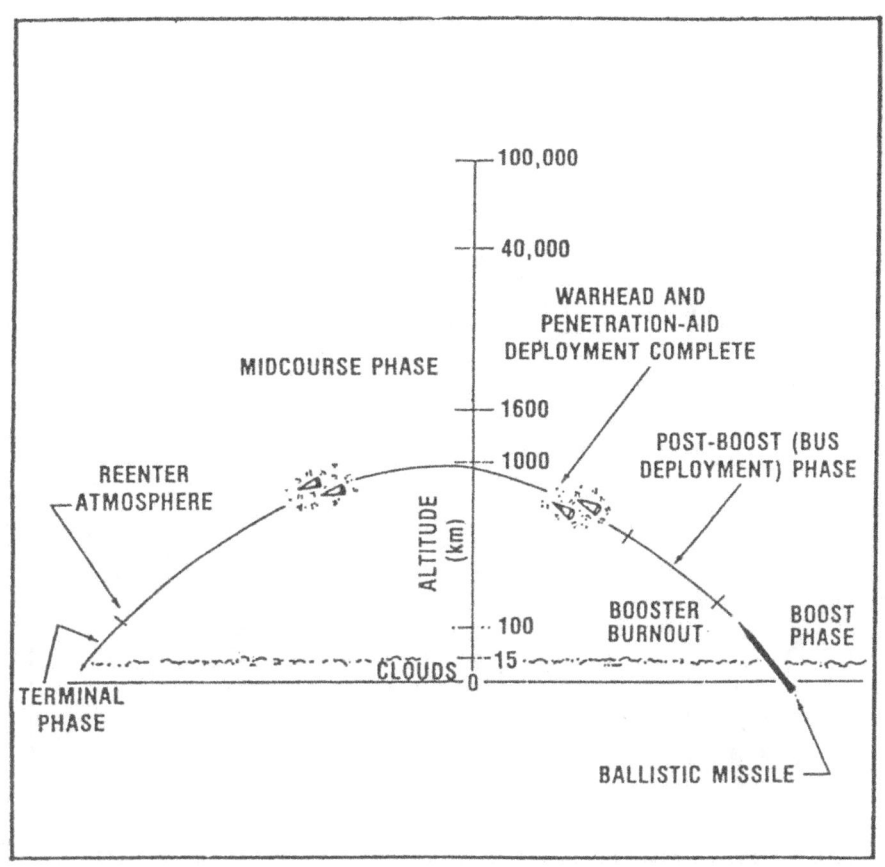

Note: During the boost phase, the rocket engines accelerate the missile payload through and out of the atmosphere and provide intense, highly specific observables. A post-boost, or bus deployment, phase occurs next, during which multiple warheads and penetration aids are released from a post-boost vehicle. In the midcourse phase, the warheads and penetration aids travel on trajectories above the atmosphere, and they reenter it in the terminal phase, where they are affected by atmospheric drag.

Source: *The Strategic Defense Initiative - Defense Technologies Study,* (Washington, D.C.: U.S. Department of Defense, April 1984), p. 14.

FIGURE 4.2 Fletcher Report: Preliminary Concept for Ballistic Missile
Defense During the Boost Phase

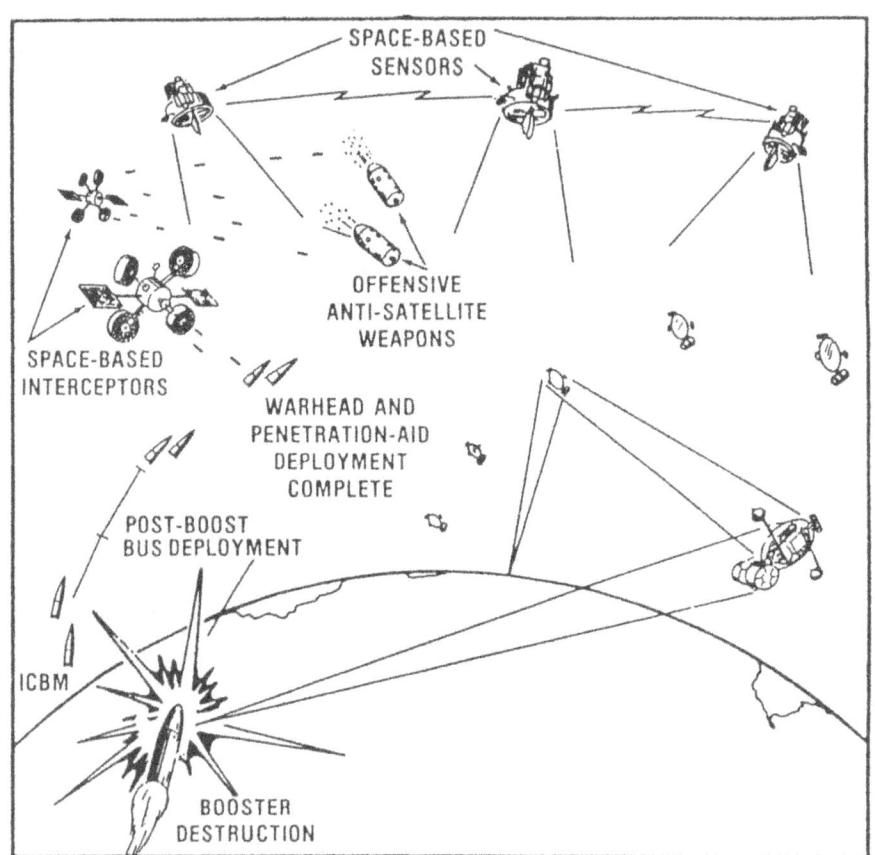

Note: An essential requirement is a global, full-time surveillance capability to detect
an attack and define its destination and intensity, determine targeted areas, and
provide data to guide boost-phase intercept and post-boost vehicle tracking systems. Attacks may range from a few missiles to a massive, simultaneous
launch. For every booster destroyed, the number of objects to be identified and
sorted out by the remaining elements of a multi-tiered defense system will be
reduced significantly and early defensive response will minimize the number of
deployed penetration aids. The transition from boost phase to midcourse
allows additional time for intercept by post-boost weapons and for discrimination between warheads and deception objects. Space-based sensors detect and
define the attack. Space-based interceptors protect the sensors from offensive
anti-satellite weapons and, as a secondary mission, attack the missiles. In this
depiction nonnuclear, direct-impact projectiles are used against the offensive
weapons.

Source: The Strategic Defense Initiative - Defense Technologies Study, (Washington,
D.C.: U.S. Department of Defense, April 1984), p. 15.

'responsive threat' problem was serious and led to a plethora of articles and commentary about potential Soviet counter-measures.[51]

Several developments helped the program: the Defense Science Board did a comparison of Soviet and U.S. space technology and concluded that American capabilities were superior in 25 of the 31 areas examined and in no area were the Soviets demonstrably ahead;[52] the Homing Overlay Experiments showed that a ground-based interceptor could hit an incoming ICBM;[53] and the Eastport study concluded that the software programming aspects of SDI were manageable as long as the architecture was not completely centralized.[54]

Nevertheless, through 1985 and into 1986, the SDIO was still relying on an expected architecture that was remarkably similar to the ideas suggested by the Fletcher Commission. In late 1985, the SDIO presented its concept to reporters and said that there would be seven independent layers of defensive weapons. This included: (a) DEWs and KKVs in the boost phase, (b) DEWs, KKVs, and ground-based lasers for the mid-course, and (c) exo- and endo-atmospheric interceptors for the terminal phase.[55] This approach to the system carried through 1986 and was still presented to the Congress in April 1987 as an example of the likely architecture. Figure 4.3 below sets out the SDIO's initial vision. This approach had some refinements, in the system components, over the Fletcher design, but it was an elaborate mixture of 3 KKV and 2 DEW systems. It would have been exceptionally expensive given the size and complexity of the space-based elements.

Considerable political capital was used to garner support for the initial SDIO architectural concept. Gen. Abrahamson spoke frequently around the country; the principal allies of the U.S. were asked to participate in joint

---

[51] See, for example, J. Caravelli, "Soviet Countermeasures to SDI," *Journal of Defense Diplomacy*, March 1985, pp. 45-47.

[52] *New York Times*, April 11, 1986, p. A-35.

[53] The validity of this test series has been hotly debated in the 1990s, as it has become known that the fourth, and only successful test, was done with a heated missile (making it easier for sensors to identify) and was hit from the side (which provided a larger target area). See, T. Weiner, "Inquiry Finds 'Star Wars' Tried Plan to Exaggerate Test Results," *New York Times*, July 23, 1994, p. 1.

[54] Cohen, D., et al., "A Report to the Director of the SDIO," Eastport Study Group, *Mimeo.*, December 1985.

[55] Mohr, C., "Antimissile Plan Seeks Thousands of Space Weapons," *New York Times*, November 3, 1985, p. A-1.

FIGURE 4.3 Example of Non-Nuclear Ground and Space-Based Architecture

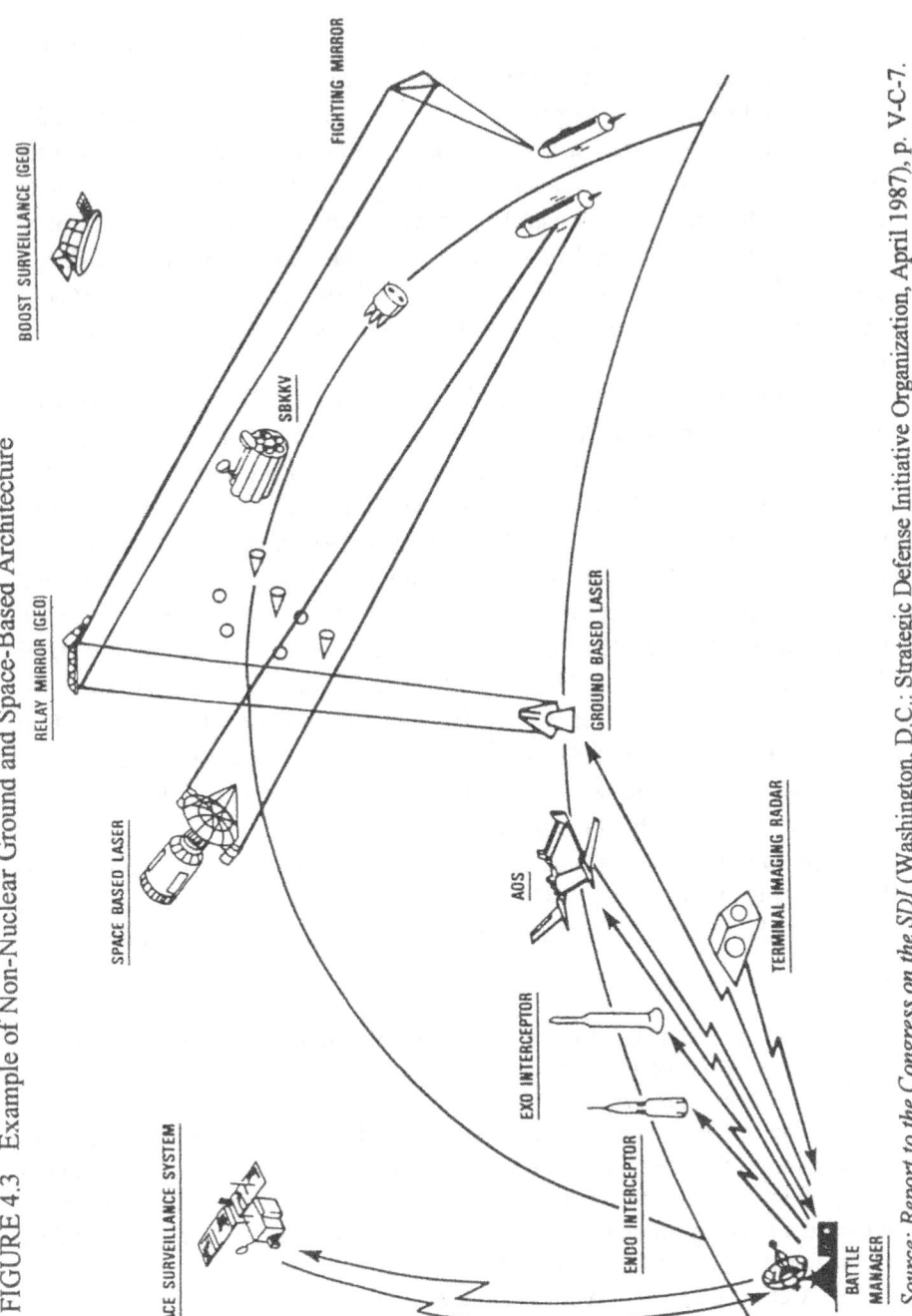

Source: Report to the Congress on the SDI (Washington, D.C.: Strategic Defense Initiative Organization, April 1987), p. V-C-7.

research efforts;[56] and a major push was undertaken to demonstrate that SDI was justified because the USSR was violating terms of the ABM Treaty.[57] However, the opponents of SDI were not resting. The Union of Concerned Scientists organized a boycott of SDI research,[58] the American Physical Society was about to release its critical report on DEWs, and the Congress was raising an increasing number of questions about the cost of the systems the SDIO was presenting.

The challenges to the early SDIO architecture led General Abrahamson to propose three important changes to the envisaged NMD system. Unfortunately, though each was technically superior and each was less expensive, the changes came in three successive years and it appeared that SDIO was frequently changing its stance. The first shift was to the Phase I Architecture, the second was to a lower cost version of Phase I, and the third was to a whole-hearted endorsement of Brilliant Pebbles. In addition to the pressure for cost reductions, SDI felt it had to fight the growing complaints of those who wanted "early deployment" and were claiming that SDI was moving too slowly.[59]

Despite having sent its April 1987 Report to Congress showing a major role for DEWs, SDIO was already planning a new system that relied exclusively on KKVs to intercept attacking missiles. The *Phase 1 Architecture* kept the Boost and Space Surveillance and Tracking Systems and Ground-Based Radars, but included only KKV Space-Based Interceptors and exo- and endo-Atmospheric Ground-Based Interceptors[60] in the plans. By dropping the DEWs (or putting them off to the long-term future once they had been tested), SDI was jettisoning the least credible part of the system.

Yet problems immediately arose. The design for the Space-Based Interceptors (SBIs)[61] was to cluster a group of 10 rockets around a "garage" where

---

[56] Weinberger, C., "Strategic Defense in Perspective," *Defense -1986*, January-February 1986, p. 6.

[57] See, for example, a widely circulated memorandum from Secretary of Defense Weinberger to President Reagan titled, "Responding to Soviet Violations Policy Study," *Mimeo.*, (Washington, D.C.: U.S. D.O.D., November 13, 1985).

[58] Charles, D., "Rise and Fall of Star Wars," *The New Scientist*, March 20, 1993. p. 24.

[59] The Early Deployment effort will be discussed later in this chapter.

[60] By this time, the preferred exo-atmospheric interceptor was ERIS, and the endo-atmospheric variant was HEDI.

[61] The term SBIs referred to the combined system of "garage" plus 10 interceptor rockets.

they could be serviced and kept ready for activation. Although the clustering was a cost-saving technique which had advantages both during launching and over the life of the SBI, it had the enormous disadvantage of potentially becoming a prime target itself.

Gen. Abrahamson originally stated that the space-based components would cost approximately $60 billion and the ground-based parts of the system about $40 billion. With great effort, the SDIO was able to get the top policy-makers on the Pentagon's Defense Acquisition Board to approve Phase I on July 30, 1987.[62] Yet, the Cost Analysis Improvement Group (CAIG)[63] said that SDIO's actual expenditures were likely to total $146 billion, (46% more than the briefings has claimed).

Although the Phase I architecture had passed an important barrier in the Department of Defense review process, the cost problems led to intense criticism both inside the Pentagon and on Capitol Hill. This produced a fevered effort inside the SDIO to cut costs; and, by June 1988, through various attempts at acquisition and manufacturing savings, the CAIG re-estimated the cost of Phase 1 at $115 billion.

Yet, like the "Perils of Pauline," once SDIO began to admit that not all of its components were necessary, further challenges to the program arose from every direction. Congress cut the Fiscal Year 1988 budget for SDIO from the $5.7 Billion requested to $3.9 billion. This forced several important delays in systems evaluation and development: (a) the number of tests for the HEDI and ERIS interceptors was reduced, (b) the assessment of the Space Surveillance and Tracking System versus ground based sensors was delayed, and (c) funds for DEWs were sharply cut.[64]

More importantly, though, there were growing doubts about whether SDIO was willing to make hard choices among the alternative systems that could be used for NMD. This led Secretary Weinberger to appoint a Defense Science Board panel Chaired by Robert Everett to do an independent assessment of SDI. This panel urged even further reductions in Gen. Abrahamson's plans. Figure 4.4 below illustrates the changing estimates for the cost of deploying the Phase 1 architecture. Not only do these estimates represent a reduction in

---

[62]   In some analyses of SDI and in SDIO documents, this was referred to as "Milestone 1."

[63]   The CAIG is an independent set of Department of Defense examiners with extensive experience in acquisition programs who evaluate the cost estimates of program managers seeking approval for purchasing particular systems.

[64]   Sanger, D., "'Star Wars' Facing Cuts and Delays - '92 Goal in Doubt," *New York Times*, November 22, 1987, p. 40.

FIGURE 4.4 Space Defense System Acquisition Cost

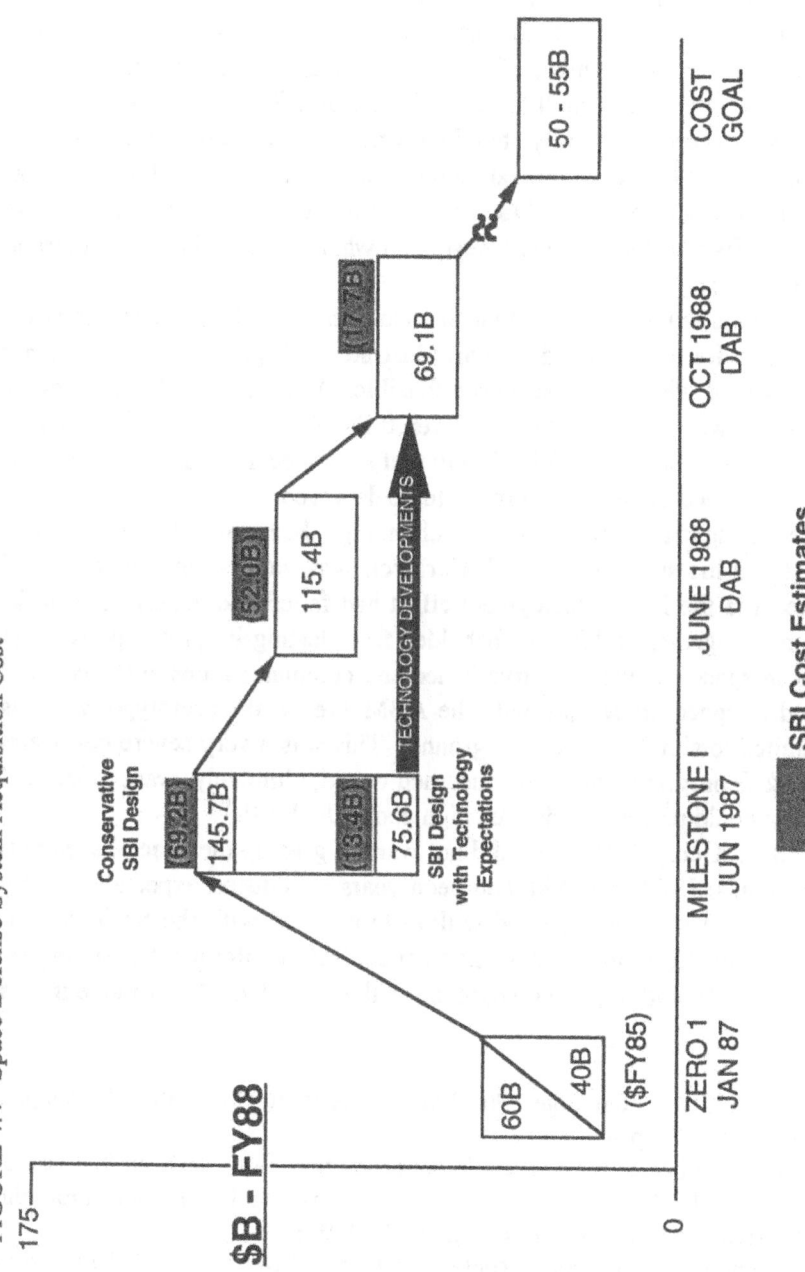

Source: SDI Briefing Charts (Washington, D.C.: Strategic Defense Initiative Organization, October 1988).

the quantity and complexity of weapons and sensors expected to be deployed in space, but a slowing of the process as well.

In May 1988 the Everett Report was completed. It recommended concentrating on the deployment of 100 ground based interceptors at Grand Forks, N.D., improving satellite surveillance, an effort to deal with Soviet SLBMs, and a phased, step by step deployment of other ground-based interceptor sites if the USSR continued violating the ABM Treaty; but the report saw the deployment of space-based systems as very far in the future.[65] In essence, the Everett Panel urged focusing on what could be done with current, proven technology.

The cost reductions between June and October of 1988 ($115 billion down to $69 billion for the deployed system) illustrated in Figure 4.4 show a further scaling back of what was considered feasible. We might call this *Phase 1 Modified*. It was also clear that the Everett Panel saw space-based weapons as such a conflict with the ABM Treaty that such a decision should be put off until fully verified systems were ready to be deployed.

The managerial setting was also changing. Secretary Weinberger had resigned and his successor, Frank Carlucci, was more skeptical of SDI.[66] Also, because the Early Deployment effort had failed and generated a backlash from Congress, SDIO was forbidden from testing its prototype weapon systems in space. Although surveillance and communications systems could be tested in space, to comply with the ABM Treaty, the prototype weapons were limited to simulations on the ground. This was a very severe constraint on an R & D program where all concerned would, ultimately, want to know if the system could perform as desired high above the Earth.[67]

Although Gen. Abrahamson did not want to give up the space-based part of SDI's plans, 1987 and 1988 had been years of reduced expectations. By September of 1988, he appeared willing to proceed with the basic recommendations of the Everett Panel and concentrate deployment planning on HEDI and ERIS and improved space surveillance systems.[68] A move in this

---

[65]   Strobel, W., "Plan Urges 'Star Wars' to Be Build Gradually," *Washington Times*, May 20, 1988, p. 4.

[66]   Although Secretary Carlucci did not concentrate particularly on SDI issues, in 1993 he revealed his concerns. See, P. Lewis, "Ex-Foes Trade Stories From the Cold War Trenches," *New York Times*, March 1, 1993, p. A-7.

[67]   Ground-based interceptors, conforming to the ABM treaty, could be tested in space.

[68]   See an extensive interview given to the press on September 7, 1988, J. Cushman, "Pentagon Official Proposes Cost Cut for Space Weapon," *New York Times*, September 8, 1988, p. A-1.

direction would at least have laid the basis for some missile defense and gotten the support of Senator Sam Nunn who had, in January 1988, called for reducing SDI to just a ground-based system, capable of protecting against accidental launches.

Nevertheless, it is clear that Gen. Abrahamson had never wanted to go in that direction; and, throughout 1988, SDIO had been working with Lawrence-Livermore labs on a proposal to radically redesign the KKVs to be put in space. Table 4.1 below illustrates how frequently the key components of the system were being changed, and the new KKVs were still another new departure. The essential idea behind the new KKVs was to make them significantly smaller and supply them with their own sensors and flight control systems so they could be dispersed in space. Since a previous KKV idea had been called Smart Rocks, these were called Brilliant Pebbles. Thus, even though the official position was to stick with Phase 1 Modified, by the time Gen. Abrahamson retired in January 1989, he was strongly urging a transition in the architecture toward *Brilliant Pebbles.*[69]

Early Deployment

The Fletcher Commission urged an extended period of R & D testing, and one of the earliest SDIO decisions was to plan on preparing alternative systems so that the President and Congress could decide the final deployment issues in the early 1990s.[70] Putting the deployment decision off for nearly a decade had several advantages for an R & D program: (a) it meant that the results of system testing would be well-advanced or completed; (b) it was beyond President Reagan's second term, so it would clearly be another Administration making the decisions; and (c) the opponents of missile defense would have a less precise target for criticism until they knew the final recommended system.

There were two principal drawbacks to this approach, however: (i) it meant that, for an additional decade, the U.S. would have no functioning NMD, and (ii) it is extremely difficult for any political leadership to

---

[69] See, for example, his "End of Tour Report" which stressed that Brilliant Pebbles were a dramatic breakthrough which could be deployed in five years and would cost a total of $25 billion. This report has not been officially released, but extensive excerpts appeared in, *Inside the Pentagon*, March 17, 1989, pp. 18-20.

[70] In various documents this was referred to as 1992, in others as "decisions to be determined in 1993."

TABLE 4.1 Evolution of Space-Based Components

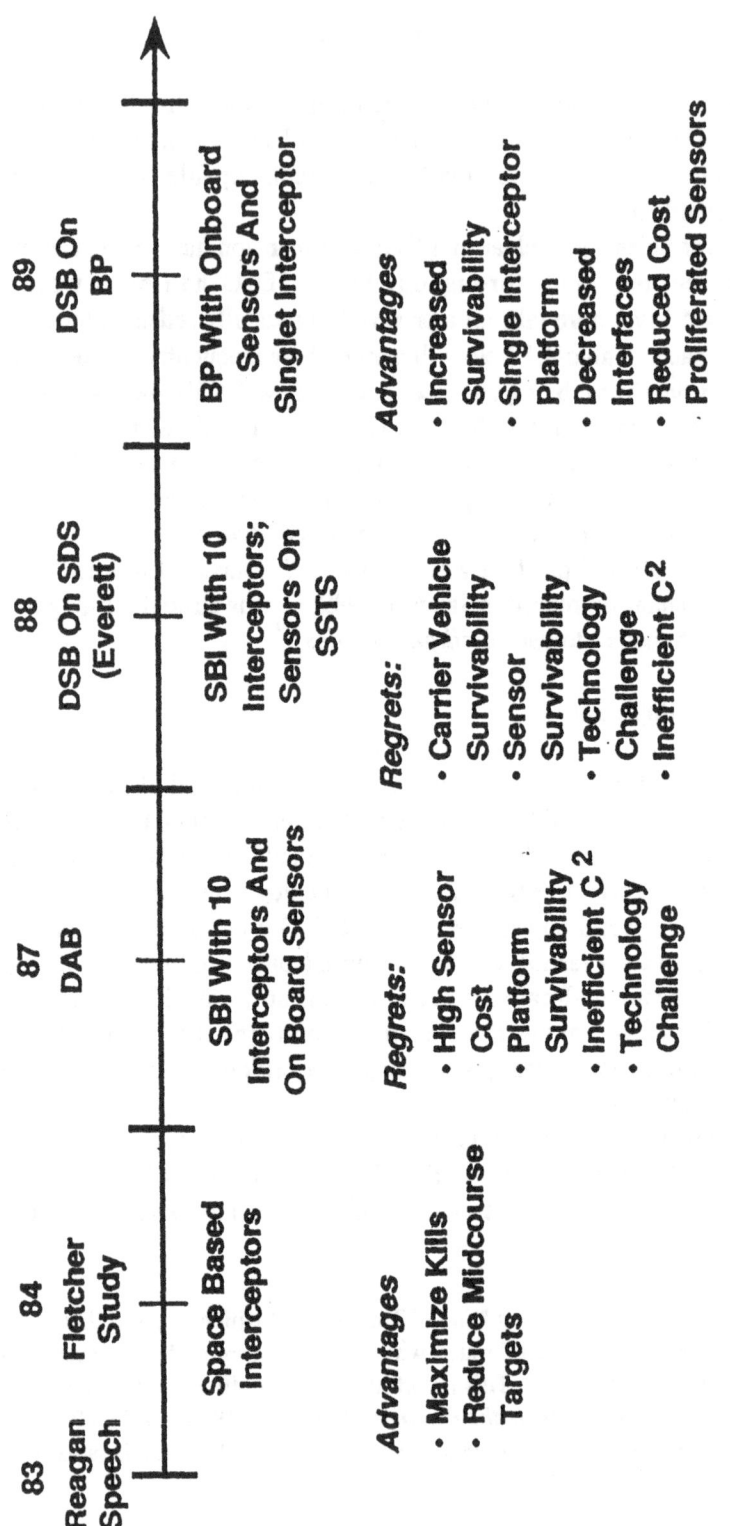

83	84	87	88	89
Reagan Speech	Fletcher Study	DAB	DSB On SDS (Everett)	DSB On BP

Space Based Interceptors

Advantages
· Maximize Kills
· Reduce Midcourse Targets

SBI With 10 Interceptors And On Board Sensors

Regrets:
· High Sensor Cost
· Platform Survivability
· Inefficient C^2
· Technology Challenge

SBI With 10 Interceptors; Sensors On SSTS

Regrets:
· Carrier Vehicle Survivability
· Sensor Survivability
· Technology Challenge
· Inefficient C^2

BP With Onboard Sensors And Singlet Interceptor

Advantages
· Increased Survivability
· Single Interceptor Platform
· Decreased Interfaces
· Reduced Cost
· Proliferated Sensors

Source: Brilliant Pebbles Program - Evolution and Current Focus Briefing (Washington, D.C.: Strategic Defense Initiative Organization, August 30, 1990).

maintain enthusiasm and active support for a program where the tangible results are more than a decade ahead.

Although the High Frontier organization, led by Gen. D. Graham (ret.) and a small group of Republican Congressmen and Senators[71] had met with President Reagan in 1985 to push for an early decision on ground-based interceptors, the Early Deployment effort did not become highly visible until two years later. It was in 1987 that Congressmen Kemp and Courter and Senator Wallop pledged to introduce legislation that would require deployment of NMD by 1993 which, because of the lag times in manufacturing, would have meant a final decision by 1988.[72]

Cong. Courter did introduce an amendment to the Fiscal Year 1988 defense authorization bill to proceed with a "first generation strategic defense" that could later be integrated into a more comprehensive NMD. The R & D Subcommittee of the House Armed Services Committee voted down the Courter Amendment 11-7 in April 1987.[73] On May 12, 1987 the full House voted down two additional amendments designed to speed the deployment decision and start initially with ground-based interceptors to deal with accidental launches.[74]

Several additional factors complicated the Early Deployment effort: on April 6, 1987 Congressman Jack Kemp announced his intention to run for President and to transform the race into "a national referendum on SDI;"[75] and the Reagan Administration was in the middle of its battle to "broadly interpret" the wording of the 1972 ABM Treaty. Moreover, Gen. Abrahamson was criticized in a Senate staff report for encouraging a reorientation of SDIO programs and "not awaiting a publicly announced Presidential decision to commit to near-term deployment."[76]

SDI had operated in a highly politicized environment from the beginning but the Early Deployment effort produced a negative reaction from the Democrats. Sam Nunn gave a series of speeches on the Senate floor rebuking the administration over the ABM Treaty interpretation, and Congressman

---

[71] The Capitol Hill group consisted of: Senators M. Wallop, P. Wilson, and D. Quayle, and the Congressmen were J. Courter and J. Kemp.

[72] Norman, C., "News and Comment," *Science*, January 16, 1987, pp. 279-280.

[73] *Congressional Quarterly*, April 4, 1987, p. 614.

[74] *Congressional Quarterly*, May 16, 1987, p. 975.

[75] Shribman, D., "Kemp Joins GOP Race for Presidency, Vows to Make It Star Wars Referendum," *Wall Street Journal*, April 7, 1987, p. 70.

[76] *SDI Monitor*, "SDI Budget Summary Reveals Near-Term Tilt," April 22, 1987, p. 1.

Aspin succeeded on April 2, 1987 in getting an 11-7 vote in the House Armed Services subcommittee in favor of the "strict" interpretation of the Treaty.[77]

The rancor that resulted from the Early Deployment effort carried over into the Fall of 1987 and was a key factor that led to the cuts in appropriations for Fiscal Year 1988 (discussed above). Hence, Early Deployment appears to have been a bad tactical error because SDIO did not yet have sufficient performance data to convince even the mildly skeptical. When the negative response led to a ban on space-based testing, it then became a real constraint on demonstrating the future performance of NMD systems.

Brilliant Pebbles

The original idea for Brilliant Pebbles (BP) was suggested, in November 1986, by Greg Canavan, a physicist from Los Alamos Labs. In a discussion with Edward Teller and Lowell Wood, Canavan expressed concern that the Space Based Interceptors (SBIs) SDIO was planning for the Phase 1 Architecture were too vulnerable and that smaller KKVs were necessary. This led, eventually, to a totally different design, relying on recent improvements in miniaturization of computers, guidance mechanisms, and sensors.

The resulting BPs were "singlets" - individual KKVs that could operate on their own. Because of their size, they would be inexpensive to manufacture and launch, so the U.S. could deploy large numbers of them in low-earth orbit. At last, it seemed the U.S. had found a missile defense technology that was demonstrably cheaper than offensive weapons.

BPs were first publicly discussed at a Washington, D.C. conference in March 1988;[78] and, it was on the basis of early tests that Gen. Abrahamson thought BPs would be a substitute for the KKVs in Phase 1.[79] However, it was not until early 1990 that the SDIO officially included BPs in its preferred architecture.

In his briefing on Fiscal Year 1991 plans, the second Director of SDIO, Lt. Gen. G. Monahan, proposed replacing the SBIs with BPs, but continued to support a large-scale, comprehensive missile defense with Boost phase, Space-based, and Ground-based Surveillance and Tracking Systems, plus Ground Based Radars and Interceptors and DEWs to follow in future years.[80]

---

[77] *Congressional Quarterly*, April 4, 1987, p. 614.

[78] Broad, W., *Teller's War*, (New York: Simon and Schuster, 1992), p. 253.

[79] For a picture of a BP, see SDIO briefing chart, # 89U-0412, (Washington, D.C.: SDIO, March 15, 1989).

[80] See, Lt. Gen. G. Monahan, *Strategic Defense Initiative - FY 1991 Program Briefing*, (Washington, D.C.: SDIO, February 2, 1990).

Figure 4.5 below gives a pictorial representation of how BPs fit into the overall architecture.

With the Berlin Wall already down and the Soviet Union in internal crisis, there were few takers for Monahan's vast enterprise. In fact, while these briefings were going on, Ambassador Henry Cooper was already working on his report which led to the down-sizing of SDI performance requirements and the shift to the Global Protection Against Limited Strikes (GPALS) system.

Although there was some modest technical criticism of BPs, from the Defense Science Board[81] and from the arms control community,[82] most observers were impressed that they were a very cost effective system. It was also clear, by the Fall of 1990, that BPs were the system of choice and that DEW research was going to be cut back drastically.[83] Although the total number of BPs planned for the GPALS system was only 1/10 as many as for Phase 1,[84] the plans for BPs survived throughout the Bush Administration as the preferred means of achieving efficiency in missile defense.

The Link Between ABM Treaty Issues and SDI

The debate about SDI and the ABM Treaty began the day after President Reagan's speech in March 1983 and it continues until today. On balance, advocates of missile defense see the Treaty as unnecessarily confining and a relic of the Cold War, where two nations were dominant and attempting to mitigate the effects of their arms competition. Supporters of the Treaty question whether missile defenses will work effectively and argue that the U.S. is still better off attempting to limit NMDs and relying on Deterrence.

There are several aspects of Treaty compliance questions which are discussed in detail in Chapter 6, but it is worth noting here that the debate over the "Broad" versus "Narrow" interpretation of the ABM Treaty had a critical effect on the support for SDI.

Although Article V of the Treaty clearly forbids the deployment of sea-

---

[81] *Aviation Week and Space Technology*, "Pentagon Science Advisers Criticize SDIO's Rush to Adopt Brilliant Pebbles," April 9, 1990, p. 23.

[82] Garwin, R., "Are Brilliant Pebbles All That Brilliant?" *Aerospace America*, December 1990, p. 6.

[83] *SDI Monitor*, "Effect of SDI Budget Cut Slowly Emerging," November 9, 1990, p. 249.

[84] The number of BPs planned for each system is still classified; but various sources have said that Phase 1 needed 4600 BPs, whereas GPALS needed approximately 1000 BPs.

FIGURE 4.5 Phase I Architecture with Brilliant Pebbles

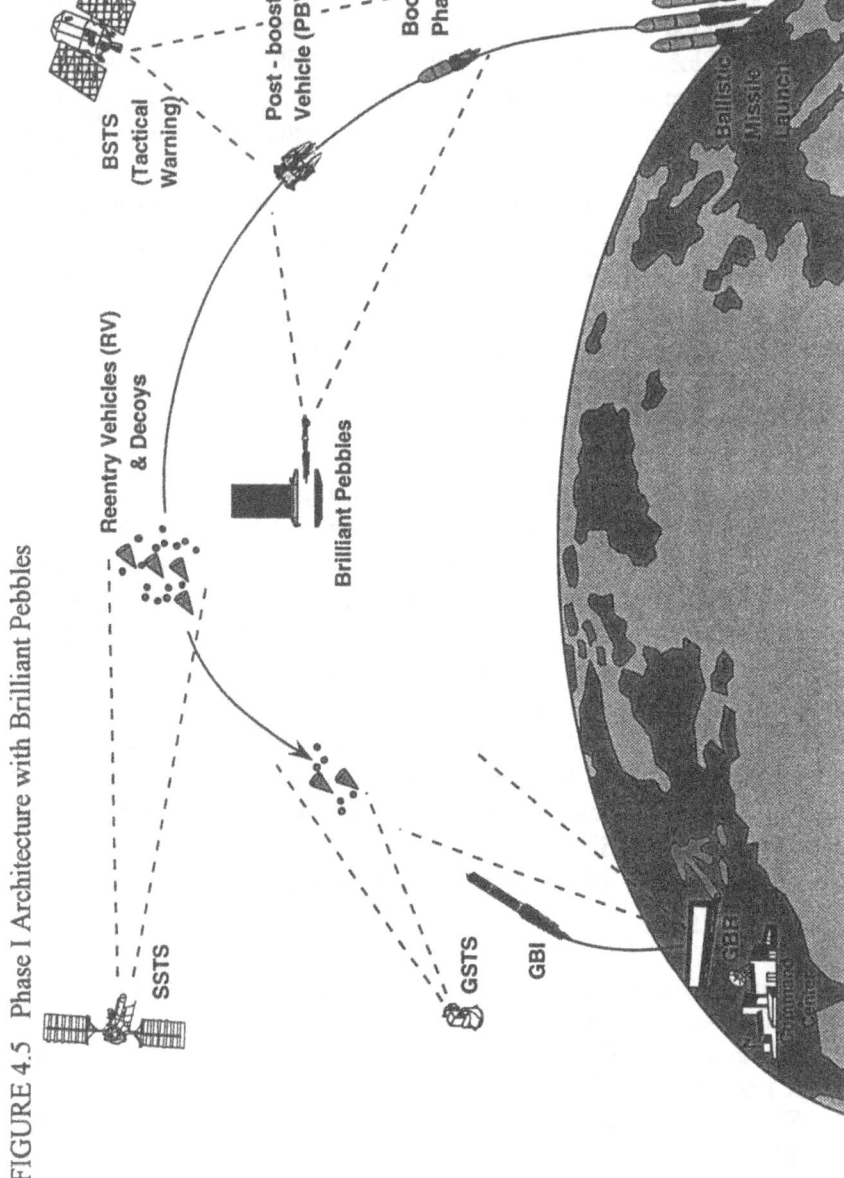

Source: Lt. Gen. G. Monahan, *Strategic Defense Initiative - FY 1991 Program Briefing*, (Washington, D.C.: Strategic Defense Initiative Organization, February 2, 1990).

based, air-based, space-based, or mobile land-based ABMs, Agreed Statement D permits testing and development of ABM systems based on "other physical principles" than those that were discussed at the time of the 1972 Treaty negotiations. Therefore, supporters of missile defense research concluded that virtually any ABM system could go through testing and development as long as it used technologies derived after 1972.

On Sunday, October 6, 1985, National Security Adviser Bud McFarlane said on "Meet the Press" that the Reagan Administration intended to adhere to the ABM Treaty but that Agreed Statement D allowed development and testing of a full range of new systems. This became known as the "Broad" interpretation of the Treaty. The "Narrow" interpretation was the one the U.S. had followed in the prior 13 years, i.e. that only fixed, land-based systems were permitted for testing.

This arcane debate generated enormous controversy because supporters of missile defense realized that they could not demonstrate the efficacy of boost phase intercepting without space-based testing.[85] Conversely, opponents realized that they could critically weaken SDI if there were no realistic tests to evaluate the new technologies. This was the real significance of Senator Nunn's speeches against the "Broad" view of the Treaty and the House Armed Services Subcommittee vote in favor of the "Narrow" interpretation.[86] Ironically, it was not until the early 1990s, when Former Soviet Foreign Minister Bessmertnykh acknowledged it was SDI which convinced the Kremlin leadership they could not compete with the U.S.,[87] that American foreign policy specialists realized how important missile defense efforts had been.

* * *

In sum, SDI went through three major changes in scale and concept before becoming GPALS and steadily lost support as its technical judgments were

---

[85] For an overview of the splits within the Reagan Administration on this issue, see, G. Shultz, *Turmoil and Triumph - My Years as Secretary of State*, (New York: Charles Scribner's Sons, 1993), pp. 578-581.

[86] There were a host of other issues involved, including: defining what "testing in an ABM mode" meant, what types of surveillance systems were permitted, and how narrowly to define what testing of "components of an ABM system" meant. For elaboration on these topics, see, S. Graybeal and P. McFate, *The ABM Treaty and Ballistic Missile Defense*, (Washington, D.C.: AAAS Publication # 93-26S, 1993).

[87] Reuter, "SDI, Chernobyl Are Said to Have Helped End Cold War," *Washington Post*, February 27, 1993, p. A-24.

challenged and the need for a comprehensive missile defense declined. Yet, it laid the basis for currently available options and could quickly be revived as a deployment program if the threat warranted it. At present, it appears that Theater Missile Defense and some type of more limited NMD will get the primary attention, but comprehensive defense could reappear like a phoenix in the 21st Century.

5

National Missile Defense

The purpose of this chapter is to show how quickly and significantly positions on National Missile Defense (NMD) changed at the end of the Cold War. The end of U.S.-Soviet tension did lead to important reductions in overall defense expenditures in both Moscow and Washington. Yet, other developments at the same time (the coup attempt against Gorbachev, friction among states of the Former Soviet Union, and the Persian Gulf War) all led to heightened support for ballistic missile defenses.

Somewhat ironically, then, just as cuts in strategic weapons were being agreed upon and as conventional forces were being reduced, BMD systems were gaining legitimacy. The passage of the Missile Defense Act of 1991 was, arguably, the all time high point of Congressional support for BMD,[1] and it came just when other national security programs were being sharply curtailed. The intent here is to explore how this situation developed and suggest reasons, as well, for the gradual erosion of support for NMD after 1991.

To put the post-Cold War period in context, it is useful to review briefly the ebb and flow of support for National Missile Defense. During the past twenty-five years there have been five principal positions regarding the desirability of NMD for the continental U.S.:

---

[1] This issue will be discussed below. The reason that some see it as more supportive of BMD than the authorization of the Safeguard system in 1969 is that the 1991 Act included a specific mandate for a thin population defense, whereas Safeguard was only designed as a point defense for missile fields and the nation's capital.

1) Those *opposing* it under all circumstances;

2) Those *opposing* NMD for most situations but willing to have a thin missile defense compliant with the ABM Treaty to deal with accidental launch and chances of a limited attack;

3) Those nominally *favoring* NMD but who see it primarily as a bargaining chip to be negotiated away for reductions in other nations offensive forces;

4) Those *favoring* NMD as a means to enhance Deterrence by protecting retaliatory forces and command and control facilities; and

5) Those *favoring* NMD for population defense.

As discussed in Chapter 2, since 1958, when the Soviet Union launched Sputnik and it became clear that ICBMs could be built, one of the most basic strategic choices facing U.S. decision-makers was whether to seek security through offensive forces (and the threat of retaliation) or whether to have active missile defenses.[2] In the 1950s and early 1960s, this was a relatively straight-forward choice because the Bradley Report demonstrated that ICBMs, even with primitive decoys, could overwhelm the defensive technologies of the period.[3]

It was not until the late 1960s, when better interceptors were available and improvements in radars had taken place, that it became feasible to destroy a larger number of incoming reentry vehicles through direct hits or fuzed conventional explosives. Hence, since the late 1960s and the wrangling over Sprint, Spartan, and Safeguard, the real debate regarding missile defense has been about strategy and the pros and cons of relying on offensive versus defensive systems. Figure 5.1 below attempts to portray graphically how U.S. presidential preferences have oscillated on this issue.

It is important to note, however, that even though missile defenses received

---

[2] After the advent of the Polaris program, there has been no opposition to "passive defenses" which seek to protect the retaliatory forces by hiding them under water or hardening land-based missile silos.

[3] At that time, the U.S. did not have fast-climbing interceptor rockets and was considering crude and large nuclear explosions in the upper atmosphere as the main means of stopping incoming ICBMs.

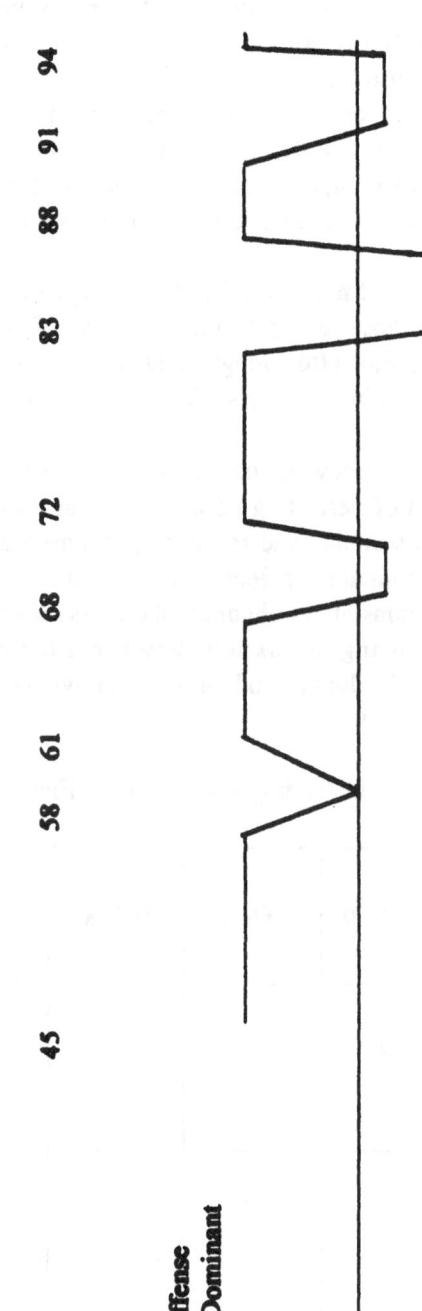

FIGURE 5.1 Preferred Presidential Policies: Offense Versus Defense

major support from Presidents Nixon, Reagan, and Bush, they have never, at any point, replaced offensive systems as the mainstay of the U.S. strategic posture. This is partly because no one has been able to devise a "leak-proof" defense. Also, because the U.S. is a major power, it has needed to maintain the ability to project force and influence (including the threat of nuclear weapons) which defenses alone cannot do. Thus, supporters of missile defense were not expecting defenses to completely replace offensive systems. The debate was usually about what mix of offensive and defensive systems to have.

Among advocates of national missile defense, there has also been a basic difference between those urging "point defenses" of missile fields and command and control facilities (to strengthen Deterrence) versus those arguing for wide coverage population defenses. Table 5.1 below depicts the dominant views on this issue.

When we note the seesaw pattern in the debate over offense versus defense and on the question of defending population versus retaliatory capability, it looks almost like a stimulus and response phenomenon. It appears that one side was able to dominate the debate for a number of years; and, then, as technology and circumstances changed, there was a shift toward a different solution. The following discussion shows that the end of the Cold War provided a remarkably fluid period in which individuals and groups changed

TABLE 5.1   BMD for Defending Population vs. Enhancing Deterrence

	1958-60	1969-72	1983-87	1988-90	1991-92
Defending Population	XX		XX		XX
Enhancing Deterrence		XX		XX	

Note: Years not mentioned were ones when there was no dominant view of the objectives for BMD.

positions frequently, and the perceived need for missile defenses fluctuated noticeably.

1988-90: NMD to Enhance Deterrence

Once the Early Deployment effort faded, supporters of NMD took a more cautious stance and there was a period of about three years when several competing views were given for continuing with NMD planning and research. One of the most comprehensive rationales was the argument for a *"cooperative transition"* from security based on offense to systems emphasizing defense.[4]

This type of transition would require both Russia and the U.S. to accept the validity of missile defenses as a stabilizing substitute for Deterrence. It would also entail agreement, beyond START II, on further offensive arms reductions and a mutually acceptable schedule for deploying and building up missile defenses. Even though there was ample evidence, by 1992, that President Yeltsin actually preferred this route, the severe economic decline of Russia has meant that it is extremely unlikely that Moscow will allocate new funds to expand missile defenses significantly. Although the Galosh system had been functioning since the late 1960s, it is clearly out of date, and a more convincing defense would require substantial expenditures. Thus, some BMD advocates hoped that Russia would endorse the "cooperative transition", but few indications that this would occur anytime soon.

On January 19, 1988 Senator Sam Nunn proposed a dramatic redirection for the Strategic Defense Initiative, urging that the Reagan Administration "take a cold shower of reality" and down-size its effort to concentrate on the limited goal of an *Accidental Launch Protection System* (ALPS).[5] Nunn's suggestion was a direct challenge to the "Phase 1" architecture proposed by Gen. Abrahamson, and stressed that SDI should concentrate on two objectives: (1) in the near-term, develop a limited system to protect against accidental and unauthorized missile launches, and (2) fund longer-term research in directed energy that "offered the best prospects for a com-

---

[4] For an elaborated version of this view, see, C.F. Ikle, "Nuclear Strategy - Can There Be a Happy Ending?" *Foreign Affairs*, Spring 1985, Vol. 63, No. 4, pp. 810-826.

[5] Nunn, S., "Nunn Outlines New Arms Control and Strategic Modernization Agenda," *Mimeo.*, (Washington, D.C.: Office of Senator Nunn, January 19, 1988), pp. 1-13.

prehensive defense." Both of these objectives could be achieved while still abiding by the terms of the 1972 ABM Treaty.

The ALPS concept was widely discussed and drew an interesting mix of supporters and opponents. Strong supporters of NMD disliked it because it undercut the basic Reagan objective of a comprehensive population defense. Ironically, it was also opposed by the arms control community because they saw it as a means for just the opposite: "the first step in the slippery slope toward SDI deployments." Pragmatic supporters of missile defense liked ALPS, however, because they saw it as a way to have some protection without the expense of SDI-Phase 1. We will see below, that the basic features of ALPS were incorporated into the Missile Defense Act of 1991. Yet, interestingly, until the coup attempt against M. Gorbachev in 1991, the sentiment against missile defense in the arms control community and in the Congress was so strong that even the limited capabilities suggested for ALPS were resisted.

During the last year of the Reagan Administration, Frank Carlucci was Secretary of Defense, and he took an eclectic approach to NMD. In his 1989 report to Congress on SDI, Carlucci strongly emphasized the continuing "Soviet challenge" and urged that the U.S. response include a combination of measures: (a) continuing offensive forces modernization, (b) a vigorous SDI program, and (c) a broad-ranging arms control effort.[6]

These steps were definitely seen in Cold War terms, and Carlucci presented the SDI program as a means to "provide a necessary and powerful deterrent to any near-term Soviet decision to expand rapidly its ABM capability beyond that permitted in the ABM Treaty."[7] It is worth noting that, in President Reagan's final year, NMD was seen as intricately linked with arms control. SDI was viewed as a complement to the START talks, and the ABM Treaty was seen as a flexible document that would permit further technical development, testing of systems "regardless of basing mode," and extensive deployment of space-based sensors.

Although "the cooperative transition," ALPS, and eclecticism were all present in the 1988-90 period, the rationale for NMD that appeared in the most visible Administration statements was to *enhance deterrence*. General

---

[6] Carlucci, F., *1989 Report to the Congress on the Strategic Defense Initiative*, (Washington, D.C.: U.S. Department of Defense, S.D.I.O., March 13, 1989), pp. 1-4 to 1-7.

[7] Carlucci, F., *Ibid.*, pp. 1-6. The same concern led early planners in the SDIO to favor having rapidly deployable ABMs if there was a significant Soviet advance (or "breakout") in BMDs or offensive missile systems.

Edward Rowny, leader of the U.S. delegation to the START talks, saw strategic defense specifically in military terms: as a way to foil any Soviet first strike and to convince Soviet strategists that attacking the U.S. would be insufficiently successful to warrant planning such aggression.[8] Rowny's vision of SDI was very much the way the Joint Chiefs of Staff conceived of missile defense. It was not seen as a leak-proof shield for population, but instead thought of as a partially-effective means to break up a cluster of incoming warheads. It was assumed that, without the certainty of destroying U.S. land-based missiles and much of the American bomber force, the USSR would see a first-strike as pointless.

In the first two years of the Bush Administration, Secretary of Defense Cheney continued to stress enhancing deterrence as the principal reason for missile defense.[9] In press briefings, Cheney said that, if the U.S. could reliably intercept 40% of the first wave of Soviet missiles and 50% of all the SS-18s, this would be sufficiently effective to make an attack by Moscow implausible. These numbers and this way of viewing the problem were almost identical to the compromise worked out between the SDIO and the Joint Chiefs in the middle-1980s.[10]

Secretary Cheney obviously had doubts about the direction SDI should take and he favored having Ambassador Henry Cooper do an independent review of the program as part of the overall Bush NSR 12 strategic assessment. Nevertheless, until Cooper's report was done, Cheney stuck to enhancing deterrence as the rationale for SDI funding.

It is clear, however, that, in the 1988-90 period, there was no senior government official pressing hard for NMD as a replacement for Deterrence. Instead, various competing justifications were extant at the same time, and NMD was seen as a supplement to other ongoing systems and initiatives. Thus, SDI was on uncertain ground. It faced three principal challenges.

The cost of the system was the most visible problem. Although the arms control community had consistently opposed NMD on principle, some of the

---

[8] Rowny, E., "SDI: Enhancing Security and Stability," *Current Policy*, (Washington, D.C.: U.S. Department of State, No. 1058, April 1988), p. 2.

[9] Cheney, R., *Report of the Secretary of Defense to the President and the Congress*, (Washington, D.C., U.S. Government Printing Office, 1990), p. 35.

[10] As discussed in Chapter IV, Gen. Abrahamson, Director of the SDIO, preferred to concentrate on population defense. The Joint Chiefs were more skeptical of SDI capabilities, but were willing to endorse the program if it had what they considered to be realistic goals. It was on this basis that goals were agreed upon for intercepting specific percentages of Soviet SS-18s and SS-19s.

most damaging early attacks on SDI had been with over-stated estimates of what it would cost to manufacture and deploy. Yet, even supporters of missile defense knew that the early architectures suggested were so vast, redundant, and expensive that they would have to be pared down. As noted in Chapter 4, one of the principal arguments in favor of Brilliant Pebbles was the reductions in cost that it was expected to achieve.

Figure 5.2 below shows the evolution of Department of Defense cost estimates for complete Phase 1 systems in the 1987-90 period, and compares that to the first estimate for the cost of the Global Protection Against Limited Strikes (GPALS) system. Not surprisingly, as the SDIO kept releasing new cost estimates, skepticism grew on two grounds: (a) Would the new system be as capable as the prior one? and (b) If such large cost reductions were feasible in a short period of time, why hadn't the potential savings been incorporated in earlier years?

These implicit criticisms were not entirely valid because the program's research was producing technological improvements that reduced hardware costs and economies of scale in procurement. Moreover, the biggest drop in estimated costs ($46.3 billion) came between June and September of 1988 when Brilliant Pebbles was officially introduced into the Phase I architecture. Nevertheless, the credibility of the overall program clearly suffered as the SDIO kept stretching to say it could defend the continental U.S. for less. By the time GPALS was introduced as a concept, the strategic environment was less threatening. There was explicit recognition that GPALS was less capable than SDI-Phase 1 and meant to deal only with a reduced threat. This led to the final $21 billion reduction in cost estimates (from $53 to $32 billion).

A second challenge that NMD efforts faced in 1989 was internal to the Executive Branch. President Bush clearly saw foreign policy as his principal interest and area of expertise. Upon entering office, he authorized two major inter-agency studies to reevaluate U.S. national security policy. These studies had begun as the situation in the Soviet Union was changing dramatically; and, by the summer of 1989, the circumstances in Eastern Europe were very volatile as well.

The Bush Administration not only wanted to analyze these changes accurately, but was very conscious about releasing results of a study that might appear to assess these global changes in too pessimistic a light. Hence, National Security Reviews 12 and 14 took over six months and, ultimately, did not provide a detailed strategic agenda. This left a controversial program like SDI in continued limbo.

Leaving NMD as just another program to enhance deterrence made it an easy target for a variety of opponents. Bruce MacDonald noted:

FIGURE 5.2 Changes in System Acquisition Cost (FY 88 $ in billions)

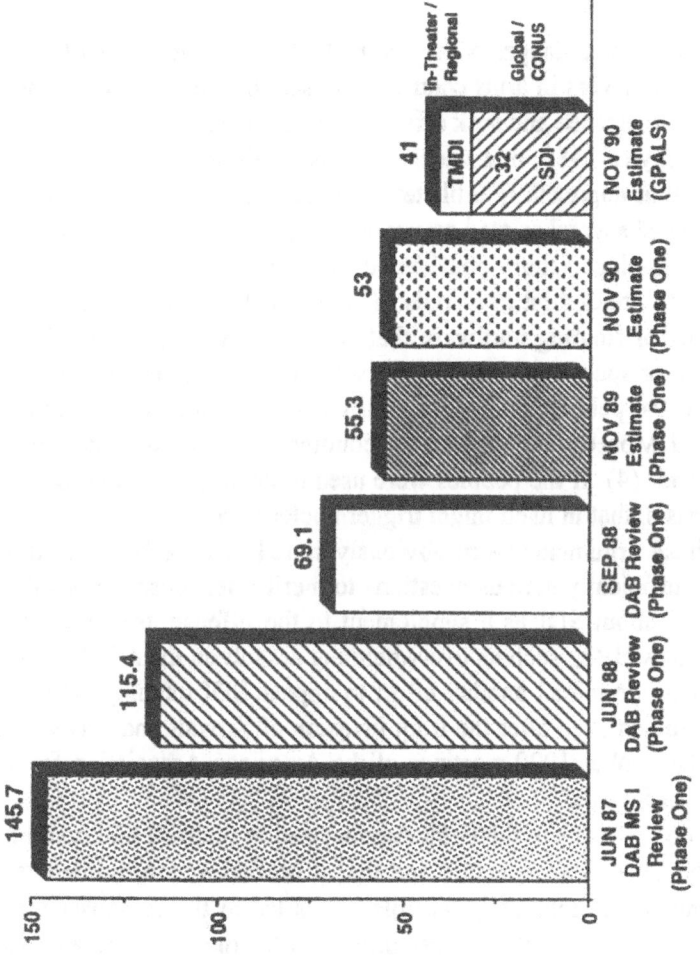

Source: Briefing on the Strategic Defense Initiative, (Washington, D.C.: SDIO, U.S. Department of Defense, February 1991), p. 12454).

> Seeking to strengthen deterrence puts SDI on a crowded playing
> field, along with the B-2 bomber, Trident II, Midgetman, and the
> whole panoply of U.S. strategic offensive forces....many of these
> other programs could do it with greater certainty at less cost, without
> destroying the ABM Treaty and thereby scuttling the prospects for
> offensive arms reductions.[11]

An additional danger of presenting NMD as a supplement to the offense is
that true believers in arms control could see the overall effort as an attempt to
discard Deterrence and seek offensive dominance.

In May of 1989, in a widely discussed editorial, *The New Republic* took
exactly that approach in a blistering attack on Brilliant Pebbles (BP).[12] The
critique had several related arguments: (1) BP could make war more likely if
the USSR placed more confidence in the pebbles than was warranted and
decided to attack first in a crisis; (2) The pebbles would pollute outer space
with lots of floating hardware that could create future difficulties for other
satellites or space stations; (3) BP was likely to trigger an arms race in space
because the pebbles could potentially shoot down Soviet surveillance satellites and Moscow would have to counter with anti-satellite weapons of its
own; and (4) If the pebbles were used to destroy Soviet space systems during a crisis, that in itself might trigger nuclear war.

These arguments were obviously stated in hyperbolic fashion but they
raised sufficiently serious questions to merit a response. In addition to these
concerns about SDI as a supplement to the offense, the arms control community saw BP as such an overwhelming assault on the ABM Treaty that they
were ready to mount another effort to oppose SDI funding. The resistance to
BP continued even after the Iraqi invasion of Kuwait and was vividly evident
at the December 1990 meetings of the American Association for the Advancement of Science.[13]

Senior officials in the Bush Administration realized SDI could not survive
with the Phase I architecture and that it needed a more limited set of objectives that were technically feasible. This led to the commissioning of Amb.
Henry Cooper's report on future directions for the SDI program, and a down-
sizing of the capabilities expected from the resulting BMDs.

---

[11] MacDonald, B., "Falling Star: SDI's Troubled Seventh Year," *Arms Control
Today,* September 1991, p. 7.

[12] *The New Republic,* "Too Brilliant By Half," editorial, May 29, 1989, pp. 7-9.

[13] See, for example, the panel discussion between Albert Carnesale and Henry
Cooper, *Mimeo.,* (Washington, D.C.: AAAS transcript, December 14, 1990).

The Influence of START and the Persian Gulf War

As the Bush Administration was laying the groundwork for a basic change in SDI, two developments fundamentally altered the context in which NMD was being assessed. Progress on the START negotiations meant there was a reasonable chance that the size of the Soviet missile force would be reduced substantially, and the Persian Gulf War proved to be the first test of whether ballistic missile defenses would work under battlefield conditions.

START[14] was critical in several regards. Most importantly, it was the first cut in strategic arms in the post-World War II period.[15] Secondly, the extent of the cut was a clear sign the USSR had given up hope of a first strike capability and it appeared to be less interested in maintaining nuclear arms as a means of threat for political objectives. Also, in purely military terms, a 50% cut in Soviet warheads meant that a U.S. NMD could be, potentially, significantly more effective should there be an actual attack on the continental U.S. Thus, there was strong evidence that the Soviet Union saw limited value in many of its offensive systems and would consider further arms reductions, even the "cooperative transition" to missile defense-oriented strategies.

Although START was not signed until July 31, 1991, its basic terms were broadly agreed upon during 1990.[16] The key features were as follows:

- within 7 years of the signing, a 50% cut in Soviet ballistic missile war- heads and a 35% cut in U.S. warheads,
- a ceiling of 1600 strategic offensive delivery systems, including silo- based ICBMs, SLBMs, heavy bombers, and mobile missile launchers,
- each party could have no more than 6000 ICBM warheads,
- no new heavy missiles, and the USSR could have only 10 warheads on each for its 154 SS-18s,

---

[14] The first arms reduction treaty was officially named, the START. Although its name never changed, when the Russians agreed to a second arms reduction treaty, for purpose of simplicity, the treaties have usually been referred to as: START I and START II.

[15] There is, of course, an elaborate debate about the usefulness of SALT I and SALT II. Both of those treaties set limits on the growth of strategic systems, but nei- ther produced actual cuts. Also, although President Carter initialed SALT II, it was never ratified by the U.S. Senate, and was only adhered to on a de facto basis.

[16] See, D. Hoffman and J. Young, "A Hallway Huddle Caps 9 Years of Arms Bargaining," *Washington Post National Weekly Edition*, July 22-28, 1991, p. 16, for a detailed summary of START provisions.

- mobile, land-based missiles must be kept within small, designated areas of 9.65 square miles,
- bombers that carry cruise missiles must be kept in separate areas from those that do carry them, and most missile-carrying bombers would be assumed capable of transporting 8-10 missiles,
- the total number of sea-launched cruise missiles would be limited to 880,
- Moscow reserved the right to withdraw from START if U.S. tests in the SDI program went beyond the provisions of the 1972 ABM Treaty.

The breadth and scope of the START agreement was a tribute to the vision and persistence of the Reagan and Bush negotiators. It reversed the direction of prior growth in nuclear arsenals and got both the U.S. and the Soviet Union to agree to discard those systems that were least essential for Deterrence. It had been under negotiation for almost a decade; and, in a certain sense, START represented the transition to the post-Cold War era.[17] Although it legally sanctioned equality in nuclear systems, START showed the weaknesses of the USSR because the terms were essentially those proposed by the U.S. Within six months after the signing, the Soviet Union no longer existed.

With these dramatic changes in the geopolitical scene, it was no longer plausible for the arms control community to claim, as it had in the 1980s, that Moscow would respond with a surge of new expenditures and systems if the U.S. began the transition to greater reliance on missile defenses.

The Persian Gulf War was also a critical turning point in the way in which the American public and, ultimately, the Congress viewed missile defense. The details of the controversy over Patriot effectiveness were reviewed in Chapter 3; but, in February 1991, Scott Armstrong succinctly captured the change in sentiment about missile defenses:

> After years of controversy and budget cuts, 'Star Wars' may be about to get a proton of respect.
> The reason is not so much the program itself, which, after seven years and $24 billion, is still far from its goal of being able to zap thousands of warheads speeding through space.

---

[17]   In the last two years of the START negotiations, the Warsaw Pact was disbanded, East and West Germany were reunited, and the Soviet Union had been forced, through economic necessity, to cut off its aid to client states like Vietnam and Cuba.

Instead, it is because of the euphoria over the success of the Patriot missile in the Persian Gulf, shifting perceptions of who America's enemies are, and President Bush's decision to narrow the goals of the program...

The result is a sharpening debate over the kind of defenses the nation needs - a debate likely to make SDI once again a central part of this year's budget dispute...[18]

Of course, what the general public did not know is that Patriot was never funded by the SDI program: it had been started 25 years before by the U.S. Army as an anti-aircraft missile[19] and only recently upgraded to have a TMD capability. Also, as covered in Chapter 4, the technical problems of dealing with ICBMs in the cold, black trajectories of outer space are far more difficult than identifying and intercepting TBMs. Nevertheless, NMD programs definitely basked in the glow of TMD successes.

There were several aspects of theater missile usage in the Persian Gulf war that will probably affect the way in which future strategists think about both TMD and NMD:

1) Missile accuracy was not critical to create fear and potential chaos among the public in the countries being attacked.
2) Finding and destroying mobile missile launchers was exceedingly difficult. There was not one confirmed success by a U.S. or allied aircraft at destroying a loaded Iraqi-Scud launcher on the ground, and allied forces had complete air-superiority by the third day of the war.
3) If TMDs are to be effective, they must be quickly deployed, and space surveillance (with real time down-links) is probably necessary to achieve satisfactory interception rates.
4) Since it is unlikely that other countries would assist significantly if the continental U.S. were threatened with missiles, the U.S. cannot rely on an allied missile defense effort. Thus, if the U.S. is to have a plausible missile defense, its surveillance and BMD systems need to be continuously active or deployable on very short notice.

---

[18]   Armstrong, S., "Patriot Missile Successes in Gulf Give 'Star Wars' New Credibility," *Christian Science Monitor*, February 6, 1991, p. 9.

[19]   The Patriot was originally named the SAM-D. Ironically, the program was given its initial impetus as an anti-aircraft missile by Secretary of Defense McNamara who became a vehement opponent of missile defenses after leaving office.

1991: Global Protection Against Limited Strikes (GPALS)

For most of the period between March 1990, when Ambassador Cooper finished his report on the future of SDI, and January 29, 1991, when President Bush announced his support for the GPALS concept in the State of the Union address, Cheney's Defense Department was preoccupied with the planning for and execution of the Persian Gulf War.

Nevertheless, the timing of President Bush's statement was opportune. The American public saw the damage that even poorly manufactured and inaccurate missiles could do and realized that imperfect defenses were better than nothing. Also, Moscow's willingness to cooperate with the allied war effort (against its former ally, Iraq) was a clear indication that the U.S.-Soviet relationship was moving into a new phase.

Ambassador Cooper's report of March 1990 is still classified but its principal recommendations are known. Cooper made four basic points: (a) the Soviet ICBM threat still existed but the chances of a large-scale first strike had been significantly reduced, (b) if there was a Soviet attack, it was more likely to be from an accidental launch or a "rogue commander" who was taking action on his own, (c) the spread of theater ballistic missiles would clearly endanger future U.S. forces deployed overseas, and (d) there was a small but growing number of third world countries that would eventually have ICBMs and they could, potentially, target the continental U.S.[20]

Cooper, thus, recommended that the theater missile defense program proceed apace but that NMD efforts be scaled back to deal with limited attacks on the U.S. of up to 200 warheads.[21] This approach: left the popular TMD programs alone, used Senator Nunn's idea of missile defense to deal with accidental launch but extended it to deal with intentional threats from the third world,[22] and down-sized the NMD goals to ones that were realistic given the technology available in the 1990s.

---

[20] For a summary of these arguments, see, H. Cooper, "End of Tour Report," *Mimeo.*, (Washington, D.C.: Strategic Defense Initiative Office, January 20, 1993), pp. 1-3.

[21] In comparison, planners for SDI-Phase 1 assumed that the USSR might be able to launch up to 2000 warheads in a first strike. The 200 figure came from estimates of the number of warheads on a Soviet missile submarine that could, conceivably, be launched by a rogue commander.

[22] Hence, one key difference between Nunn's idea of an Accidental Launch Protection System (ALPS) and GPALS was that the latter was intended to deal with *intentional* as well as accidental missile firings. A second difference was that

The essential features of Cooper's report were incorporated into the GPALS initiative. The day after President Bush's speech, the Department of Defense issued a statement that summarized the key elements of the GPALS program:

> ... The primary responsibility for maintaining deterrence of an intentional, massive Soviet offensive strike against the U.S. and its allies will remain the U.S. strategic offensive forces for the remainder of the century. The purpose of GPALS is to provide protection against limited strikes.
>
> A GPALS defensive system would consist of the following:
>
> - Space and surface based sensors to provide global continuous surveillance and tracking, from launch to intercept or impact, of ballistic missiles of all ranges...
> - Interceptors, based both in space and on the ground or at sea, capable of high-confidence protection to targets under attack. Space-based interceptors could provide continuous, global interdiction capability against missiles with ranges in excess of 600-800 kilometers. The surface-based interceptors, located in the U.S., deployed with U.S. forces, and potentially deployed by U.S. allies, could intercept missiles of any range and with any type of warhead. Interceptors would use non-nuclear, hit-to-kill technology.[23]

On February 12, 1991, Assistant Secretary of Defense Stephen Hadley and Ambassador Cooper (who, by then, was Director of the SDIO) gave a briefing about GPALS that was broadcast on CSPAN.[24] The briefing emphasized that GPALS would have a smaller scale and lower budget than SDI-Phase 1 and would not rely on any futuristic Directed Energy Weapons. Cooper also pointed out that the miniaturized sensors planned, Brilliant Eyes, could be used to greatly improve the performance of TMDs. For example, by giving the precise coordinates of an incoming missile like a CSS-2, the same ground-based TMD could fire sooner and therefore protect significantly more area.

(Continued)

GPALS included Brilliant Pebbles and ALPS did not.

[23] Press release, "New Strategic Defense Initiative Focus: Global Protection Against Limited Strikes (GPALS)," *Mimeo.*, (Washington, D.C.: Office of the Assistant Secretary of Defense for Public Affairs, January 30, 1991), p. 2.

[24] For a detailed summary of the briefing, see, H. Cooper and S. Hadley, "Briefing on the Refocused Strategic Defense Initiative (edited transcript)," *Mimeo.*, (Washington, D.C.: U.S. Department of Defense, February 12, 1991), 75 pp.

If the GPALS program was fully funded, the Department of Defense leadership anticipated that modernized TMDs (Patriot PAC-3, ERINT, and THAAD) could be deployed by the mid-1990s. The ground-based interceptors for ICBMs and the space-based Brilliant Pebbles and surveillance satellites (Brilliant Eyes) would be available for deployment between 1999 and 2001.[25]

The initial reaction to GPALS was mixed. Technical specialists were impressed that the SDIO was now supporting a system that appeared feasible and effective, albeit for a reduced threat. The response from arms control advocates was, not surprisingly, vehement opposition. Although there were few claims that this system would not work, it was resisted by arms controllers on two grounds: (a) the space-based Brilliant Pebbles were a violation of the ABM Treaty and would lead to either a renegotiation of the Treaty or a collapse of the entire Treaty regime, and (b) the proposed ground-based interceptors, capable of handling incoming missiles of all ranges, would also be a violation of the ABM Treaty if they were shared with allies or deployed at more than one site.

Conversely, the strongest supporters of missile defense were disappointed because they saw GPALS as a weak compromise which would delay deployment of the ground-based interceptors until the end of the century. Frank Gaffney, for example, feared the Bush Administration would not push hard enough for Brilliant Pebbles and that this would mean the U.S. would go throughout the 1990s without any substantial missile defense.[26]

Senator Nunn was skeptical of Brilliant Pebbles, but favored the new focus on TMD systems and thought it might be feasible to renegotiate the ABM Treaty with the Soviet Union to get up to 6 sites for ground-based interceptors.

All of this maneuvering changed dramatically, however, after the coup attempt against President Gorbachev in August 1991. Not only was the Soviet Union on the verge of disintegrating into the Commonwealth of Independent States, but it now appeared, for a brief period at least, that the leadership in Moscow had uncertain control over the military.[27] The specter

---

[25] Cooper, H. and S. Hadley, *Ibid.*, p. 14322a.

[26] Gaffney, F. Jr., "Hapless SDI Damsel in Distress," *The Washington Times*, June 10, 1991, p. D-3.

[27] Interestingly, even a die-hard opponent of SDI like Henry Kendall, a prominent member of the Union of Concerned Scientists, had earlier seen the issue of accidental nuclear war as troubling. See, B. Blair and H. Kendall, "Accidental Nuclear War," *Scientific American*, December 1990, Vol. 263, No. 6, pp. 53-58.

of Belarus, Ukraine, and Kazakhstan having their own ICBMs and instability within Russia concentrated attention on and generated support for missile defense. It also showed that the assumptions beyond Mutual Assured Destruction were as inapplicable to an unstable industrial country as they were to a third world nation like Iraq.

The Congress then began to view missile defense as a pressing issue. After an upsetting trip to Moscow, Senators Warner, Lugar, and Cohen submitted a bill calling for deployment of 1200 ground-based interceptors in the continental U.S.; Senator Nunn began to mobilize support for a variant of ALPS; and Les Aspin, then Chairman of the House Armed Services Committee, published a very pessimistic assessment of the situation in the Soviet Union.[28] By November of 1991 both Houses of Congress were sufficiently concerned about missile defense that majorities were found to support the Missile Defense Act of 1991.[29] Table 5.2 below outlines the key provisions of the bill.

In a very important sense, the Missile Defense Act represents the high-water mark of support for BMD in the U.S. It specifically authorized: (i) deployment of a ground-based interceptor (GBI) battery at one site, (ii) planning for the construction of additional GBI sites if the USSR agreed to modifying the ABM Treaty, (iii) space-based sensors, (iv) modernized TMDs, (v) continued research on Brilliant Pebbles, and (vi) urged the President to negotiate various additional changes in the ABM Treaty to make the operation of BMDs more effective, yet still Treaty-compliant.

Had the Congress provided adequate funding for the bill and not trimmed back its scope in future years, the Missile Defense Act might have laid the groundwork for an important change in U.S. strategy. Yet, in subsequent years, the Congressional support for the GBIs waned and NMD has become solely an R & D program.

Many strong supporters of BMD, inside the Bush Administration and out, felt that a very important opportunity was lost at the time when the Missile Defense Act was being drafted. There was a chance of getting the Congress to authorize deployment (not just R & D money) for Brilliant Pebbles. Had this occurred, the U.S. would have been required to fundamentally renegotiate or abandon the ABM Treaty. True believers in Deterrence saw this as heresy and claimed that the Soviet Union would never agree. However, as we will

---

[28] Aspin, L., "A New Kind of Threat - Nuclear Weapons in an Uncertain Soviet Union," (Washington, D.C.: House Armed Services Committee, September 12, 1991).

[29] The Act was signed into law on December 5, 1991.

TABLE 5.2 Provisions of the Missile Defense Act of 1991

Goals
- Deploy an ABM system, including one or an adequate additional number of ABM sites and space-based sensors, that is capable of providing a highly effective defense of the United States against limited attacks of Ballistic Missiles
- Maintain strategic stability
- Provide highly effective Theater Missile Defenses to forward deployed and expeditionary elements of U.S. Armed Forces and to U.S. friends and allies

Theater Missile Defenses (TMD)
- Aggressively pursue the development of advanced Theater Missile Defense systems with the objective of down selecting and deploying such systems by the mid-1990s
- Development of deployable and rapidly relocatable advanced Theater Missile Defenses capable of defending forward-deployed and expeditionary elements of the Armed Forces of the United States
- Cooperation with friendly and allied nations in the development of Theater Defenses against Tactical or Theater Ballistic Missiles

Initial Deployment
- Develop for deployment by the earliest date allowed by the availability of appropriate technology or by FY 96 a cost effective, operationally effective, and ABM Treaty-compliant ABM system at a single site as the initial step toward deployment of the ABM system described in the first goal listed above
  - 100 Ground-based Interceptors (the design of which will be determined by competition and down selection)
  - Fixed, ground-based, ABM Battle Management Radars
  - Optimum utilization or space sensors including sensors capable of cuing Ground-based ABM interceptors and providing initial targeting vectors

Limited Defense System (NMD)
- Development of systems, components and architectures for a deployable ABM system capable of providing a highly effective defense of the U.S. against limited strikes, but below a threshold that would bring into question strategic stability
  - Includes activities necessary to develop and test systems, components, and architectures capable of deployment by FY 96 as part of an ABM Treaty-compliant initial site defense system
  - For purpose of planning, evaluation, design, and effectiveness studies, such programs, projects, and activities may take into consideration both the current limitations of the ABM Treaty and modest changes to its numerical limitations and its limitations on the use of Space-based Sensors

TABLE 5.2 (Continued)

Space-Based Interceptors (GMD)
- Conduct research on Space-based Kinetic-Kill Interceptors and associated Sensors that could provide an overlay to Ground-based ABM Interceptors
- Robust funding for research and development, for follow-on technologies, including Brilliant Pebbles, is required
- Deployment of Brilliant Pebbles is not included in the initial plan for the limited defense system architecture
- Report on conceptual and burden sharing issues associated with the option of deploying Space-based Interceptors (including Brilliant Pebbles) for the purpose of providing global defenses against Ballistic Missile attacks

ABM Treaty Negotiations
- Congress recognizes the President's call for "immediate" concrete steps to permit the deployment of defenses against limited Ballistic Missile strikes and the Soviets undertaking to consider such proposals from the United States on non-nuclear ABM systems
- Congress urges the President to pursue immediate discussions with the Soviets on the feasibility and mutual interests of amendments to the ABM Treaty to permit
  - Additional ground sites and interceptors
  - Increased use of space sensors for direct battle management
  - Clarification of development and testing
  - Flexibility for advanced ABM technology
  - Clarification between TMD and ABM defenses

Review of Deployment Options

- Interim report due May 94 on progress of negotiations
- Assess progress and consider options to the U.S. as now exist under the ABM Treaty

Deployment Plan

- Within 180 days, submit deployment plan for TMD systems and the ABM system established by the the goals of the 1991 Missile Defense Act

NMD=National Missile Defense
GMD=Global Missile Defense
TMD=Theater Missile Defense

Source: *Report to Congress: Plan for Deployment of Theater and National Ballistic Missile Defenses*, (Washington, D.C.: U.S. Department of Defense, June 1992), p. 2.

see below, within two months, President Yeltsin had given a speech where he called for joint U.S.-Soviet cooperation on a global missile defense system. Thus, it is certainly possible that, if the Bush Administration had been bolder on this issue and pressed the Congress harder, the U.S. today would have statutory authority for a "cooperative transition" toward missile defense.

START II and the "Global Protection System"

The collapse of the Soviet Union and the transition from Gorbachev to Yeltsin were probably the two most important strategic developments of the early 1990s. Neither of the changes was well-anticipated,[30] and each, on its own, was critical for the course of future U.S. policy.

The transition from Gorbachev to Yeltsin was not just a shift from Communist rule to democracy. Far more importantly, it involved the complete discrediting of the Communist Party of the Soviet Union and the gradual dismantling of the police state. Not only did this create freer expression of views and make the government turn its attention to economic performance, but the new openness in society made it far easier for outsiders to judge developments inside the country. Russia is still the world's second most powerful nation. Yet, even if Moscow were to embark on a course of military rearmament and renewed expansion, outsiders would now be far more able to judge what was occurring and the extent of the threat this posed. This is particularly important in missile defense because, on several occasions during the Cold War, the U.S. overcompensated when it was not sure of Soviet offensive capabilities and intentions.[31]

The break-up of the Soviet Union was important in strategic terms for two reasons: (a) it meant that Moscow would have more limited resources to draw upon, and (b) Russia would have to devote effort and funds to dealing with the former republics on its periphery. This resource constraint would have been critical even if Russia stayed Communist; however, now it clearly means that, for a sizable period, Russia can threaten with nuclear weapons but cannot project its power in the way that it did in the 1970s and 1980s.

---

[30] For a detailed analysis of why the research community failed to anticipate the collapse of the USSR and the subsequent demise of the Communist Party, see "The Strange Death of Soviet Communism," colloquium in *The National Interest,*Spring 1993. No. 31, pp. 10-109.

[31] Two examples of this were: (a) President Kennedy's decision to proceed with quick deployment of ICBMs even though he had found out his campaign rhetoric about "missile gaps" was wrong, and (b) President Nixon's decision to proceed with MIRVing the Minuteman missiles as the Soviet ICBM force grew.

As the Bush Administration began its planning for 1992, both the foreign policy and domestic scene looked promising. President Bush had high public popularity ratings, those perceived as his major Democratic opponents had declined to enter the presidential race, and the transformation of Eastern Europe and the Soviet Union had immensely benefited the U.S. and occurred without major bloodshed.

President Yeltsin surprised most observers by his speech at the UN on January 31, 1992 when he called for a Global Protection System. Yeltsin advocated multilateral ballistic missile defense efforts and a "global system for protection of the world community based on a reorientation of the U.S. SDI to make use of high technologies developed in Russia's defense complex."

Senior policy-makers in the Bush Administration knew that they could press for major initiatives with Russia, but they had to decide which would be given priority. There were two schools of thought about how to proceed: one group wanted to press vigorously to start the Global Protection System, get Russian assent to Brilliant Pebbles, and fundamentally redraft the ABM Treaty. The second group urged going slowly, concentrating on further reductions in offensive arms (START II), and not pushing Yeltsin too hard.

There was a whole series of judgments that entered into this decision. There were debates about political stability in Russia, the extent of Yeltsin's support, and the balance the U.S. should try to achieve in bringing about economic and defense policy changes in Moscow. It was also clear that a START II agreement was achievable within months,[32] whereas the initiation of the Global Protection System (GPS) and renegotiation of the ABM Treaty would be a far longer undertaking.

Ultimately, President Bush personally decided START II would be given priority, but talks should proceed on both the GPS and minor modifications of the ABM Treaty. The START II negotiations went rapidly, as expected, and by mid-June 1992 Boris Yeltsin came to Washington to initial the preliminary agreement. Tables 5.3 and 5.4 below show how the START limits affect different strategic systems. For example, if all goes as planned, by the year 2003, the U.S. and Russia will have reduced warhead levels from over 11,000 each to approximately 3500 each.

---

[32] The U.S. also faced a curious anomaly with START. The START Agreement had been signed with the Soviet Union which, by 1992, no longer existed. Although Russia said that it would abide by the terms, Moscow could not bind Belarus, Ukraine, and Kazakhstan. Subsequent negotiations were necessary to deal with this.

One of the most critical features, from the U.S. perspective, was the Russian concession to destroy all its "heavy missiles" (the SS-18s and SS-19s) capable of carrying a large number of MIRVed warheads.[33] The related provision, that both sides will deploy only missiles with one warhead, is stabilizing[34] and cost saving because the U.S. Minuteman IIIs are nearing obsolescence and the U.S. can simply field a new, 1-warhead missile rather than going through the interim step of manufacturing and maintaining unwanted M-Xs.[35]

There have been three principal problems that have delayed both the ratification and the implementation of the START agreements: (1) strong nationalists in both Russia and the former republics are resisting the arms re-

TABLE 5.3 Comparison of Warhead Levels Before and After START I & II

	Before START		START I		START II	
Weapons	U.S.	Russia	U.S.	Russia	U.S.	Russia
ICBM Warheads	2,450	6,595	1,444	2,800	500	705
SLBM Warheads	4,032	2,760	3,456	2,096	1,728	1,712
Bomber Weapons	4,980	1,804	4,524	1,804	1,268	1,040
Totals	11,462	11,259	9,424	6,700	3,496	3,457

Source: A. Woolf, "The START and START II Arms Control Treaties: Background and Issues," *Mimeo.*, (Washington, D.C.: Congressional Research Service, June 30, 1993), p. 9.

---

[33] The SS-18s were considered particularly threatening because they had so much thrust that Russia could have continued to add MIRVs as it mastered further miniaturization techniques. Also, it would have been extremely difficult to monitor compliance had there not been agreement to do away with the missiles entirely.

[34] MIRVed missiles have always been the subject of considerable controversy because they give the attacker the potential chance to destroy several of the opponent's missiles for each one fired. Thus, in a crisis, there might be an incentive to attack first if one side thought it was going to lose its retaliatory capability.

[35] The M-Xs are MIRVed and can carry up to 10 warheads per missile.

TABLE 5.4 Key Limited Systems Under START I and II

Weapon System	START I	START II
Total Delivery Vehicles	1,600	No Limit Specified
Warheads Attributed to all Delivery Vehicles	6,000	3,000-3,500
Warheads Attributed to all Ballistic Missiles	4,900	No Limit Specified
Warheads Attributed to MIRVed ICBMs	No Limit Specified	0
Warheads Attributed to Heavy ICBMs	1,540	0
Warheads Attributed to Mobile ICBMs	1,100	No Limit Specified
Warheads Attributed to SLBMs	No Limit Specified	1,750
Warheads Attributed to Heavy Bombers	No Limit Specified	No Limit Specified

Source: A. Woolf, "The START and START II Arms Control Treaties: Background and Issues," *Mimeo.*, (Washington, D.C.: Congressional Research Service, June 30, 1993), p. 29.

ductions; (2) it is expensive and requires highly specialized technical skills to deactivate nuclear weapons and dispose of the waste products safely; and (3) political leaders in the Ukraine distrust Russian intentions and see giving up their nuclear weapons as the loss of an important bargaining chip.[36]

---

[36] For an overview of these problems and how they have slowed the START ratification process, see, T. Friedman, "U.S.-Russia Accord on Arms Hits Snag,"

Despite these difficulties and the prospect that Russia and the former republics might not implement the understandings in a timely fashion, the Senate went ahead and ratified the START I Treaty during the middle of the presidential campaign on October 1, 1992.[37] Just before he left office, President Bush went to Moscow and on January 10, 1993 signed the final version of the START II Treaty. On balance, most observers see the START agreements as major advances and, in many ways, they can be viewed as the crowning achievements of the Bush Administration nuclear policy.

Giving priority to the START process, however, meant that there was little progress during 1992 on missile defenses and the Global Protection System. In the Cheney Defense Department, the GPS was seen as a broad concept which, at a minimum, would link the major industrial democracies in efforts to slow the spread of ballistic missiles and provide an initial defense against them. The principal, initial goal with the GPS was to set up an Early Warning Center that would consolidate surveillance data and make it available to all countries that were willing to adhere to the guidelines of the system. There was, also, a strong U.S. desire to strengthen the Missile Technology Control Regime (MTCR) to limit the flow of key hardware and components to aggressive states. Although Brilliant Pebbles could, potentially, have provided the actual means for intercepting ICBMs, neither the principal U.S. allies nor Russia were anxious to see the U.S. deploy and have sole control over a BP system. Bush Administration policy-makers might have been willing to share control of BP, but only if there was broad-ranging cooperation among all GPS member nations.

When President Yeltsin came to Washington in June 1992 for the initialing of the preliminary START II agreement, the U.S. and Russia agreed on a joint communique regarding the GPS. The communique stressed the following three steps:

- The potential for sharing Early Warning Information,
- Cooperation in developing ballistic missile defense capabilities and technologies, and
- Development of a legal basis for cooperation, including new treaties to implement a Global Protection System.[38]

(Continued)
New York Times, October 15, 1992, p. A-3.

[37] For details of the ratification vote, see, J. Cushman, "Senate Endorses Pact to Reduce Strategic Arms," *New York Times*, October 2, 1992, p. A-6.

[38] "U.S. / Russian Joint Statement on a Global Protection System," *Mimeo.*, (Washington, D.C.: Office of the Secretary of Defense, June 16, 1992), 2 pp. with notes for press guidance.

To explore how Russia and the U.S. could cooperate on the GPS, a series of unpublicized talks were arranged in the summer and early fall of 1992 between Secretary of State Baker's aide, Dennis Ross, and General Mamadoff of the Russian General Staff.

The Mamadoff / Ross Talks did not have any substantive results for several reasons: (a) President Bush's popularity ratings led most observers to believe he would not be re-elected and the Russians did not want to make concessions in the fall of 1992 if they were going to have to deal with a new administration; and (b) the Russian military saw no reason to endorse BP quickly.

Brilliant Pebbles faced a variety of challenges. On Capital Hill, the Congressional Budget Office issued a critique of the Defense Department's plans for implementing the Missile Defense Act, and recommended cutting funds for both ground-based interceptors and BP.[39] Even within the Pentagon, the Office of Program Evaluation and Analysis issued a report questioning the SDIO's procurement schedule for BP and saying that it would be unrealistic to deploy the system without many years of extensive testing.[40]

America's principal allies had real concerns as well. The Europeans were worried that U.S.-Russian negotiations on GPS might limit their role, while the Japanese had reservations about weapons in space and whether they were constitutionally free to join an alliance like GPS.[41] Finally, both Russia and the allies recognized that letting the U.S. deploy BP would give Washington strategic dominance for the next generation. Although GPALS was a "light" system designed to deal only with limited strikes, if the U.S. could deploy 1,000 BPs, there was no inherent reason why it could not deploy the 4,600 BPs originally proposed by Gen. Abrahamson.[42]

---

[39] Schmitt, E., "Agency Proposes Options to Cut 'Star Wars' Costs," *New York Times*, May 28, 1992, p. A-17.

[40] Broad, W., "Pentagon Analyst Questions Timing for Missile Shield - Crippling Problems Seen," *New York Times*, June 2, 1992, p. A-1.

[41] For an overview of U.S. discussion with the allies on GPALS and BP, see, S. Hadley, "U.S. Ballistic Missile Defense Policy," Statement Before the Senate Armed Services Committee, Subcommittee on Strategic Forces and Nuclear Deterrence, May 20, 1992.

[42] General Abrahamson originally thought that BP could do the entire job of strategic defense. He envisaged deploying as many as 10,000 individual BP singlets. By 1988, General Abrahamson favored combining BP with ground-based interceptors and was planning on 4,600 BPs.

This left GPALS and the GPS in a tenuous position at the end of the Bush Administration. Some argued that the U.S. had consistently under-estimated Moscow's reluctance to shift toward missile defense,[43] while advocates of Deterrence claimed that it was U.S. offensive systems that kept the Persian Gulf War from escalating further.[44] Overall, however, the 1992 focus on the START process meant that missile defense, especially NMD, lost much of the momentum gained during 1991. President Clinton's interests were in such totally different areas, and his campaign pledges were for Deterrence not GPALS, so it was hardly surprising that, after January 20, 1993, missile defense lost its executive branch support.

Clinton Positions

During his run for office, and in numerous speeches, President Clinton emphasized that his principal concerns were with domestic policy. His main 1993 efforts dealt with budget policy and raising taxes, and his 1994 priority has clearly been health care.

To the extent that national security has received attention, Clinton has focused on three main objectives: (1) down-sizing the military, (2) cooperating with international organizations, principally the UN, plus the use of U.S. forces for humanitarian purposes, and (3) development of a new arms control and non-proliferation policy.

Neither of the Clinton Administration's top three national security priorities has a direct bearing on NMD, but, interestingly, each has had an indirect effect. The military down-sizing, discussed in Chapter 1 of this book, has made defense funds extremely scarce; and that, in turn, makes it particularly hard to initiate any new programs.

The use of U.S. military forces for humanitarian purposes was initially popular when President Bush decided to send troops to Somalia; but it has become more and more questionable as the Bosnia and Haiti situations have evolved. These humanitarian efforts have, thus, used up Presidential support and now limit what can be done in other foreign policy arenas.

---

[43] For an interesting early variant of this argument, see, B. Lambeth and K. Lewis, *The SDI and Soviet Planning and Policy*, (Santa Monica, Cal.: RAND, No. R-3550AF, January 1988).

[44] For an elaboration of the position that it was the U.S. nuclear threat that kept Saddam Hussein from using chemical and biological weapons, see, D. Ignatius, "Who Says Nuclear Deterrence Has Ceased to Work?" *International Herald Tribune*, May 13, 1992, p. 6.

Clinton's arms control policy could, conceivably, be directly linked to NMD, however, to date, it has been something of an enigma. In July of 1993, the Administration leaked plans of a major future initiative to start a worldwide ban on the production of enriched uranium and plutonium,[45] but little came of this effort.[46] Moreover, in March of 1994, the new Secretary of Defense, William Perry, announced that the U.S. would consider the use of nuclear weapons to deter or destroy chemical and biological weapons. If the U.S. reserved this right, it seems unlikely that other countries would think having their own nuclear programs is inappropriate.[47]

The Clinton Administration position, *supporting NMD research and development but not deployment*, is based, as discussed above, on the assumption that no new power will be able to directly target the continental U.S. with ICBMs during the 1990s.[48] Former Secretary of Defense Aspin accepted that assessment and argued, in the "Bottom-Up Review," that the principal focus of BMD efforts should be on TMDs, while maintaining a technology program for NMD. Secretary Perry, his successor, reaffirmed this policy. By dramatically cutting back the Bush Administration's plans for NMD, the Clinton budget request for combined NMD and TMD programs was reduced from $39 billion to $17 billion for the Fiscal Year 1995-99 period.[49]

The U.S. Senate has been more supportive of eventual ground-based interceptor deployment than the Clinton Administration, and included in its 1994 Defense Appropriations Bill specific instructions that the Ballistic Missile Defense Office (BMDO) continue with detailed plans for GBI development.

In response to this Senate request, Lt. Gen. Malcolm O'Neill, the current Director of the BMDO, wrote a detailed letter to Senator Daniel Inouye, Chairman of the Defense Appropriations Subcommittee, summarizing the cur-

---

[45] The Bush Administration formally announced in 1992 that the U.S. would no longer produce plutonium or uranium for nuclear warheads, and this was merely making a stated policy of an informal ban that had been operating for many years.

[46] See, M. Gordon, "U.S. Hopes to Curb A-Arms By Restricting Fuel Output," *New York Times*, July 28, 1993, p. A-2.

[47] See the *New York Times* editorial on this subject, "Mr. Perry's Backward Nuclear Policy," March 24, 1994, p. A-22.

[48] The current debate over the potential range and speed of development of North Korea's Taepodong II missile could change this assessment, but the Clinton Administration is relying on the intelligence community's current forecast.

[49] Aspin, L., "The Bottom-Up Review," (Washington, D.C.: U.S. Department of Defense, September 1, 1993), p. 23.

rent administration plans for NMD.[50] O'Neill stressed that the Clinton program was "re-orienting the NMD program to a Technology Readiness Program." In the appendixes, O'Neill included a chart with three "epochs" of GBI development anticipated between 1994 and 2003. The apparent intent is to move from a long-wave infrared seeker to a multi-mode seeker and to concentrate on miniaturization and improved maneuverability for the GBI rocket. Obviously, this is a major step back from GPALS and the idea of shifting to a multilateral missile defense effort like the Global Protection System.

*   *   *

In Chapter 6, we will review some of the current public debates about missile defense as well as suggesting a non-partisan, analytical framework in which pros and cons of expenditures on TMD and NMD can be evaluated.

---

[50] O'Neill, M., "Letter to Sen. Inouye on the GBI Program," (Washington, D.C.: BMDO, U.S. Department of Defense, February 18, 1994), 4 pp. plus appendixes.

6

Missile Defense and U.S. Security Policy

Ballistic Missile Defense: How Much Is It Worth?

Much of the debate on whether to deploy ballistic missile defenses has been emotional and ideological. Although Theater Missile Defense (TMD) controversies have generally been over technical questions and uncertainty about performance in battlefield situations, the debates about National Missile Defense (NMD) have been almost theological in character.

The NMD dialogues have often been framed as: defending the integrity of the ABM Treaty regime versus recognizing the need to defend people instead of retaliatory forces. Neither of these formulations is adequate and neither is a good guide to choices that need to be made in the 1990s.

Until recently, TMD was seen as so different in scale and range from NMD that advocates of Deterrence and protecting the ABM Treaty gave it scant attention. TMD was the subject of low-keyed discussions about potential speed, accuracy, and success at intercepting different types of incoming missiles and reentry vehicles. NMD debates have a much longer history, and have been so fundamentally colored by differing views about how to deal with the Soviet Union, that the protagonists have rarely been willing to discuss the subject in incremental terms.

Two developments have blurred the distinction between TMD and NMD: (a) the increased speed of intermediate range missiles means that defenders may need to have interceptors for theater purposes that can also handle certain slower-moving ICBMs, and (b) the public perception of the Patriots' success in the Gulf War has led to greater Congressional willingness to fund missile defenses.

Thus, TMD is now receiving greater challenges from the traditional opponents of NMD. Yet, concurrently, there have been shifts in the other direction. Some former advocates of pure Deterrence are now saying that the world has changed sufficiently so that missile defenses are more necessary and there may even be merit in negotiating changes in the ABM Treaty.

There is a curious mixture of old and new in the present ballistic missile defense (BMD) debates. Most of the issues being assessed now have already been identified by previous analysts of the problems involved. For example, General Joseph Stillwell in 1946, William Bradley in 1958, and Jerome Weisner in 1962 each discussed the difficulty of discriminating between decoys and warheads.

In addition, numerous analysts have pointed out the advantages of stopping ICBMs during their Boost Phase, and it is now being recognized that the same approach may be necessary for effective defense against theater and inter-mediate range missiles. Likewise, even the issue of the number of sites for a U.S. NMD has been argued for over 25 years. The ABM Treaty initially allowed two sites, then a renegotiation took place to limit the U.S. and USSR to one each, and now there is the possibility of reopening the issue and poten-tially having as many as four or five sites.

Given the long-standing rifts over these questions, there is a strong tendency for the principal advocates and opponents to stick with their previous positions, even when the technology involved and the security environment has changed dramatically. The basic thrust of this volume is meant to be a corrective to the politicization of BMD issues and an effort at demonstrating that the topic needs pragmatic, not ideological decision-making. Several themes have been touched upon from various perspectives:

1) BMD should be viewed as a means to limit damage and as a partially effective form of insurance, not as a perfect solution to vulnerability.

2) The U.S. has long spent a sizable part of its strategic forces budget on passive defense: maintaining submarine launched ballistic missiles and hardening ICBM silos and command and control facilities, for the purpose of threatening opponents with a retaliatory strike.

3) There is no logical reason why active BMD should be seen as more threatening than passive defenses unless they become so perfect that the U.S. could attack an opponent and be fully confident of intercept-ing all incoming missiles.

4) Since perfect missile defenses are extremely unlikely and near-perfect defenses at least a generation away, the relevant question to ask is: how much should a country pay for improved levels of protection?

5) If the BMD problem is cast in these terms, it becomes a manageable benefit / cost issue not a matter of devotion to irreconcilable principles.

The Table 6.1 below lays out BMD policy choices in benefit / cost terms, and is meant to show which BMD systems would respond to which types of threats. The judgments summarized in Table 6.1 are obviously over-simplifications and would not hold true in all circumstances. For example, if an opponent was able to interfere with U.S. surveillance or command and control facilities, the success rate for intercepting incoming missiles would go down. Conversely, if U.S. intelligence efforts were able to downgrade an opponent's launch capabilities or identify launch locations immediately, then the chances of interception might be higher than the table suggests. The purpose of this table, however, is not a specific forecast but more to set out what tasks TMD and NMD might reasonably be expected to accomplish.

TABLE 6.1   BMD as a Cost-Benefit Problem

	Area of Operation	Probability of Missile Attack	Potential Damage if Not Defended	Probability of Intercepting RVs
TMD	U.S. Forces deployed overseas	High	Low, if conventional munitions; High, if chemical, biological, or nuclear weapons	Good, if not a massive one-time strike
NMD	Continental U.S.	Very low	High, if chemical or biological weapons; Catastrophic, if nuclear weapons	Good, if limited attack; Moderate, if less than massive attack

Ballistic Missile Defense as Catastrophic Insurance

Most of the debate over BMD has been qualitative, not quantitative,[1] and the positions taken were usually highly correlated with respective views about how to deal with the Cold War. Those favoring Deterrence have frequently overestimated the costs of BMD and overstated the likelihood that U.S. opponents would develop counter-measures that would effectively penetrate BMDs. Likewise, many advocates for BMD have over-promised on its effectiveness and not fully addressed the potentially destabilizing impact that might have resulted if the U.S. had both large offenses and convincing defenses at the same time.

Now that the Cold War is past, there is no plausible opponent who could immediately counter U.S. BMDs. Therefore, in thinking through whether it is wise to have BMD, we now have two advantages: (a) this can be addressed in a non-ideological fashion, and (b) we can assume limited but not full-scale responses to U.S. actions. The first step is to provide a framework of deciding if BMD is worthwhile. Table 6.2 below outlines six potential U.S. options regarding BMD.

TABLE 6.2   Options for BMD Decisions

	Zero Defense	Limited Defense	Comprehensive Defense
TMD		Current U.S. Position	
NMD	Current U.S. Position		

---

[1]   One quantitative exception to that pattern in the analyses of BMD was, B. Blechman and V. Utgoff, "The Macroeconomics of Strategic Defense," *International Security*, Winter 1986-87, Vol. 4, No. 3, pp. 33-70.

At present, the U.S. has limited TMDs with its Patriot PAC-2 batteries and the Clinton Administration has endorsed the Reagan and Bush initiatives for expanding both the number and type of TMD systems. However, the U.S. has no NMD and the Clinton Administration has avoided making any recommendation about deployment of an NMD system. What the Clinton Administration has done is to state that it will adhere to the terms of the ABM Treaty; so, if there was a decision to deploy an NMD, it would either have to be at one site or the U.S. would need to negotiate with the Russians on modifying the Treaty to permit multiple sites.

The most concise, comprehensive set of criteria stating how the U.S. Government would evaluate strategic defense (NMD) came from Paul Nitze in a speech on February 20, 1985.[2] The *Nitze Criteria* were that NMD deployments should not be made unless the system would be demonstrated to be: (1) operationally effective against a full-scale Soviet attack, (2) survivable against direct attack itself, and (3) cost-effective at the margin.

Some observers viewed the Nitze Criteria as "a way to kill SDI,"[3] while others saw it more as an effort to delay SDI while negotiations were underway on Intermediate Nuclear Forces and START I. The key criterion was the third, and its intent was to require that U.S. expenditures on NMD not be foiled by comparable offensive investments by the Soviet Union. At the time, many critics of SDI were claiming that it would be easy for the USSR to just add offensive capability (decoys and new missiles) and thus overwhelm the defensive missiles.

Neither Criterion 1 nor 3 is relevant today. Criterion 1 is outmoded because Russia appears extremely unlikely to launch a massive strike on the U.S. Therefore, designing a system to deal with that eventuality would be clearly over-spending.

Criterion 3 had a key analytic flaw even when it was published, and its weakness is even more pronounced today. There is no reason why one nation which may be wealthier, have a higher per capita income, and be more advanced technologically should limit itself to being more efficient in all its investments than its opponents. In fact, it might be perfectly rational for one country to *outspend* its opponent by a considerable margin as long as the

---

[2] See P. Nitze, "SDI, Arms Control, and Stability: Towards a New Synthesis," *U.S. Department of State, Current Policy No. 845,* (Washington, D.C., GPO, June 1986). This was the position of Mr. Nitze and Secretary of State Schultz, but it was not accepted by many at the NSC staff or in the Defense Department. President Reagan appears not to have taken a formal position on this.

[3] Talbott, S., *Master of the Game,* (New York: Vintage Books, 1988), p. 218.

wealthier country was convinced that the insurance being bought was worth more than the cost.

Hence, there are four criteria suggested here for deciding whether BMDs are worthwhile, and they apply to both TMD and NMD:

1) Will the system enhance overall strategic stability?
2) Is the system survivable against direct attack by plausible opponents?
3) Do the BMD systems give the U.S. greater flexibility in its foreign policy and defense options? and
4) Is the net present value of the protection provided by the BMDs worth more than the cost of the systems?

Entire volumes have been written on the subject of *strategic stability*, so this will not be addressed in depth here. Yet, at the most basic level, one can ask: Would the U.S. deployment of BMDs encourage others to proliferate offenses or attack in a preemptive fashion? or Would the BMDs make opponents less likely to attack?

The answers to these questions would depend on whether the opponents were prone to aggression, technically capable of overwhelming the U.S. BMDs, and willing to incur retaliatory offensive strikes from the U.S. These are empirical issues which cannot be answered in the abstract. However, the current world geopolitical scene is obviously radically different from the 1980s, where some of the leadership in the USSR might have interpreted U.S. missile defenses as potentially threatening and thus destabilizing.

Although North Korea and Iraq today may see U.S. BMD as threatening (in conjunction with U.S. offensive capability), it would be stretching credulity to argue that their responses could be destabilizing in the same way that Soviet responses might have been in the mid-1980s.

The second criterion suggested here, *survivability of the BMD*, is the same as Nitze's and is not controversial. Clearly, if the system itself is vulnerable to sabotage, software problems, or direct attack, it cannot function. To proceed with deploying an ABM that was not survivable would be a dangerous bluff.

The third aspect of deciding upon BMD deployment is whether it increases *flexibility* in U.S. policy options. One of the most disturbing features of Deterrence as a doctrine is that it is based on the ability to retaliate and could potentially cede control of a crisis to a more aggressive opponent. Hence, one of the key advantages of having both TMD and NMD is that it could increase the precision of action in difficult circumstances. With effective BMDs in place, the U.S. could plan for intervention or to increase its force presence in a regional conflict while limiting the risk of escalation. Under Deterrence, in a conflict like the 1973 Middle East War, the U.S. had to threaten nuclear escalation as a way to show its degree of commitment. With credible BMDs, the U.S. could enhance its flexibility by treating each potential conflict on its own terms.

The *present value* of the system is the fourth criterion suggested and is designed to illustrate the principal economic factors in deciding upon whether BMDs are a useful investment. The present value estimates are not meant to be the dominant criterion, nor are they able to capture many of the complexities of the ultimate decision-making process the U.S. should consider. Moreover, it should be recognized that present value calculations are static indicators and based on a large number of challengeable assumptions.[4]

Nevertheless, the analysis below is intended to lay out the main variables that Americans should consider in evaluating the economic aspects of BMD. The present value equation cannot deal with the psychological, political, or foreign policy features of the decision, but it can let readers see what the likely returns would be to an investment in BMD. The assumptions and calculations are presented below so that readers can adjust the variables if they wish to consider different threats, deployment costs, or effectiveness levels.

The basic equation proposed for estimating the economic value of a BMD system is:

$$PV = \frac{PBMA \times ED \times \frac{BSA}{C}}{(1+r)^n}$$

Where:

PV	=	present value
PBMA	=	probability of missile attack
ED	=	effectiveness of the defense
BSA	=	benefit of stopping the attack
C	=	life cycle cost of the system
r	=	social rate of discount
n	=	number of years system operates

Since the probability of a theater missile attack on U.S. forces or our allies

---

[4] Present value calculations are a means for comparing the stream of net benefits that an investment will generate vs. society's opportunity cost of using the funds in some other way.

is high and the benefits of stopping the attack are large, there has been a strong political consensus in the U.S. for developing effective TMDs. Since the probability of a missile attack on the U.S. homeland is much lower and the likelihood of developing an effective defense is also lower then for TMDs, there has been more skepticism about NMD.

Yet, to show how such an equation might be used, four scenarios for NMD in the U.S. are outlined, with estimates made of the economic value of a BMD system in each case. A number of simplifying assumptions were made:

1) The full, life-cycle cost of an NMD system of 100 launchers, 100 interceptors, ground-based radars, battle management and communication systems and personnel is $20 billion.

2) The benefits of stopping an attack are estimated as the physical (non-human) assets that would have been lost if the missiles were not stopped.

3) Human suffering would be enormous if a major missile attack occurred on U.S. urban areas. Also, the loss of human capital would be staggering. Thus, by excluding human losses here, we are substantially under-estimating societal losses. This was done because there is a major debate about how to measure human capital[5] and no agreement on how to measure suffering.

4) Estimates of the U.S. physical capital stock are based on John Kendrick's 1976 work.[6]

5) An undefended attack on the U.S. was assumed to entail:
   - 10 missiles with nuclear warheads launched,
   - 2 missiles malfunction,
   - 8 warheads land on the largest U.S. urban areas, since these are presumably the largest cities, 1/2 of the country's urban infrastructure is destroyed,
   - 1/3 of total physical stock is destroyed or useless due to radiation.

---

[5]   See, D. Jorgensen and B. Fraumeni, "Investment in Education and U.S. Economic Growth," *Mimeo.*, (Cambridge, Mass.: 1991). As this draft article demonstrates, human capital has been the largest single explanatory factor in U.S. economic growth in the 1973-1990 period. Nevertheless, estimates of the quantity of human capital vary enormously (see Table 11-B, p. 26).

[6]   Kendrick, J., *The Formation and Stocks of Total Capital*, (New York: Columbia University Press for the National Bureau of Economic Research, 1976), Table B-14, p. 193.

6) Kendrick's estimate of 1969 total capital stock was $10.9 trillion, with $3.9 trillion in human capital.

7) If we assume that it takes 5 years to build the NMD, the physical capital stock should be calculated from 1999. Thus, we have taken his 1969 estimate of $7.0 trillion in physical capital and assumed a 2% growth rate for 30 years (i.e. multiplied it by 1.812) yielding an estimate of $12.68 trillion. If 1/3 of this was destroyed in an undefended attack, the losses would be $4.227 trillion.

8) The social rate of discount is assumed to be the 30 year bond rate, i.e. 7%.

9) The chances of an attack are assumed to be equal in any one year and the total probability is cumulative; thus, the five year probability is 5 times the one year probability. Each scenario will be estimated for year 5 and year 10.

Scenario 1

- Chances of an attack are 1 out of 10.
- The system is 90% effective against the eight warheads that make it toward U.S. targets.

$$PV(5) = \frac{.5 \times .9 \times \frac{\$4.227\text{tr.}}{20}}{(1.07)^5 = 1.40} = \$67.93\text{bn.}$$

$$PV(10) = \frac{1.0 \times .9 \times \$211.35\text{bn.}}{(1.07)^{10} = 1.97} = \$96.56\text{bn.}$$

Scenario 2

- Chances of an attack are 1 out of 20.
- The effectiveness of the BMD system is 90%.

$$PV(5) = \frac{.25 \times .9 \times \$211.35\text{bn.}}{(1.40)} = \$33.97\text{bn.}$$

$$PV(10) = \frac{.5 \times .9 \times \$211.35bn.}{(1.97)} = \$48.28bn.$$

Scenario 3

- Chances of an attack are 1 out of 100
- The effectiveness of the BMD system is 90%.

$$PV(5) = \frac{.05 \times .9 \times \$211.35bn.}{(1.40)} = \$6.79bn.$$

$$PV(10) = \frac{.1 \times .9 \times \$211.35bn.}{(1.97)} = \$9.66bn.$$

Scenario 4

- Chances of an attack are 1 out of 100.
- The effectiveness of the BMD system is 99%.

$$PV(5) = \frac{.05 \times .99 \times \$211.35bn.}{(1.40)} = \$7.47bn.$$

$$PV(10) = \frac{.1 \times .99 \times \$211.35bn.}{(1.97)} = \$10.62bn.$$

It should be stressed again that these calculations significantly understate the potential benefits of a BMD system because the cost of human suffering and capital are excluded. Also, the principal purpose is to show how BMD can be viewed as insurance, yielding different returns under different circumstances. These are simulations not forecasts. They should be viewed as a means to understanding the service being purchased, not as providing a sole, quantitative rule about whether to deploy a BMD.

Nevertheless, several interesting findings result from the calculations:

- If the chances of an attack are as high as 1 out of 10 and the system functions well (i.e. at 90% effectiveness), then the system easily pays for itself after several years.

- If the chances of an attack are as low as 1 out of 100, then the system does not yield an adequate return in either 5 or 10 years.
- Although going from 90 to 99% effectiveness increases the value of the system, it is not necessary to have the system operate at the higher level to make it worthwhile as long as the risk of attack is substantial. In a low threat environment, the probability of attack is more critical than small improvements in system effectiveness.

Without going into the sample calculations, it is easy to see why there is a broad consensus of support for TMD yet, at the same time, real debate about the merits of NMD for the continental U.S. Since many countries have short to medium range missiles, the chances of U.S. troops or ships deployed overseas coming under missile attack are high and the benefits of protecting them are large.

Although the damage estimates from conventional attacks by theater missiles are, obviously, lower than for nuclear attacks on the U.S., the probability of TBM warfare occurring is so much higher that some insurance against them is clearly prudent. Moreover, it should be pointed out that "theater" attacks on U.S. forces overseas could well be seen as "strategic" by the country where the troops are operating. Therefore, many U.S. allies may see TMD in the same light as Americans see NMD.[7]

One final point on these present value calculations is that they are static and assume constant effectiveness ratios throughout the deployment of the BMDs. A technically competent and aggressive opponent would almost certainly make efforts to reduce that effectiveness. Thus, future and more sophisticated efforts at estimating the value of BMDs would need to account for dynamic changes in the capabilities of potential opponents.

With this basic analytic approach to BMD, we now turn to three key questions:

A) Should the U.S. expand its current TMD capability?
B) Should the U.S. have an NMD system? and
C) If the U.S. deploys increasingly capable BMD systems, what related

---

[7] If the U.S. and its allies both had TMDs deployed in the same theater, it would clearly improve the overall rate of effectiveness as long as the systems did not interfere with each other.

changes in policy are necessary to take advantage of this new direction in strategy?

Should the U.S. Expand and Upgrade Its TMD Capabilities?

At the time this manuscript is being drafted, there is a growing consensus in the U.S. foreign policy and defense community that expanded TMD capabilities are an important part of future American security strategy. Effective TMDs may also be important for building military coalitions among allies who would be unwilling to commit troops unless they felt missile defenses would be available to limit loss of lives and military assets. However, consensus does not necessarily imply good policy, and support for new TMDs is by no means unanimous.

Key parts of the arms control community are vehemently opposed to the newer TMD systems, just as they oppose any NMD. Spurgeon Keeny, Executive Director of the Arms control Association, set out several arguments against increasing the quantity and sophistication of TMDs:

> As the centerpiece of its "counter-proliferation" plan, the Clinton Administration has proposed a $25 billion anti-ballistic missile system to defend U.S. troops abroad and our allies from missile attack...
> To build support for the defense system - which would include 3000 land-based anti-ballistic missiles - the Administration has concocted a highly misleading litany of national security threats...
> To deploy the system, we would have to drastically modify the 1972 Antiballistic Missile Treaty. This would create a gaping loophole, permitting the deployment of systems with substantial strategic capabilities and inviting the Russians to deploy their own defense system...
> What should the world conclude about U.S. priorities when it realizes that over the next 20 years, U.S. spending on the proposed missile system would be 100 times greater than our begrudging contribution to the International Atomic Energy Agency's nonproliferation inspections?[8]

In essence, the opponents make four points: (1) the TBM threat is overstated, (2) the cost of TMD is too high, (3) TMD risks undercutting the ABM Treaty,[9] and (4) the U.S. should rely more on international inspec-

---

[8] Keeny, S., "Inventing An Enemy," *New York Times*, June 18, 1994, p. 21.

[9] This is a different version of an earlier argument that some in the arms control community made which was that all theater missile defense violated the ABM

tions rather than active defenses. To address these claims and provide an overview of the pros and cons regarding upgrading TMDs, three basic questions need to be asked:

- Is the TBM threat real?
- Are the current and anticipated future TMD systems concentrating on the right technical improvements? and
- Will the upgraded TMD systems be able to intercept ICBMs and thus blur the distinction between "theater" and "strategic" missile defense?

Is there a tangible TBM threat to U.S. forces?

As discussed in Chapter 3, in addition to Central Intelligence Agency and Defense Intelligence Agency studies, there have been two main reviews of the TBM threat: the System Planning Corporation study directed by Dr. Nagler, and the report of the International Study Group on Proliferation and Missile Defense.[10] All of these studies have concluded that the TBM threat is real as long as the U.S. intends to remain actively engaged in the Middle East and Asia.

Altogether there are over 30 different types of TBMs in LDC government inventories today. Although Libya has only a small number of ballistic missiles, Syria, Iraq, Iran, China and North Korea are all involved in active efforts to expand the number and capabilities of their TBMs. China has been exporting its M-9s and CSS-2s, both intermediate range missiles; while the North Koreans have not only been exporting the Rodong I, but have been developing the Taepodong I and II which are believed to be a series of 5000-10,000 km range missiles capable of hitting most targets in East Asia.

The Missile Technology Control Regime (MTCR) was set up in 1987 and now has 18 member countries which have agreed not to export missiles with payloads of more than 500 kg and ranges of more than 300 km. Although the MTCR has laudable goals, it is only a protocol not a legally binding treaty, has no enforcement capability, and no verification provisions. Moreover,

---

(Continued)
Treaty. See, for example, Paul Warnke's statement on NBC Nightly News on March 11, 1983 that upgrading Patriot to handle Soviet SS-20s in Europe "would be a flat violation of the ABM Treaty."

[10] Nagler, R., *Ballistic Missile Proliferation - An Emerging Threat*, (Arlington, Va.: System Planning Corporation, 1992), and Adm. J. Betermier, et. al., *Proliferation and Missile Defenses*, (Fairfax, Va.: Study Group on Proliferation and Missile Defense, NIPP, June 1993).

many of the countries it is designed to inhibit, have refused to join.[11] Thus, the MTCR may be seen as a desirable effort, but it is hardly a credible way to deal with missile proliferation today.

It is worth noting that there are some very sound, technical reasons why LDC governments prefer missiles as a way to project their power. Missiles are cheaper, easier to maintain, and require far less training for crews than medium-range aircraft. Ballistic missiles also have a devastating psychological effect as British authorities found out when the German V-2s started hitting London and as the Iranian leadership discovered when the Iraqi Scuds landed in Tehran. Even if the MTCR does become more effective in the future, it is not likely to thwart the most aggressive and innovative governments that seek expanded missile capabilities.

The recent experience in dealing with North Korea's truculence regarding its compliance with the Nuclear Non-Proliferation Treaty should also provide a cautionary warning about the effectiveness of the International Atomic Energy Agency (IAEA). Although the IAEA has a technically competent staff and offered useful warnings about North Korea's behavior, it has little direct influence and has to rely on the press and its member countries to bring real pressure to bear on a recalcitrant state. It is also possible that the October 1994 bilateral agreement between the U.S. and North Korea will encourage other potential nuclear states to circumvent the IAEA and bargain directly with the major powers for concessions.

Thus, unless the United States intends to dramatically reduce its presence in the Middle East and Asia, the threat of TBMs will be a major issue in the coming decade. So, much like the governments in Russia, Japan, South Korea, and Israel, the U.S. will probably proceed with expanding and upgrading its theater missile defenses.

Is the U.S. concentrating on the right technologies for its TMDs?

The Ballistic Missile Defense Organization (BMDO) is currently planning for a two-tiered TMD.[12] The upper-tier would have missiles that could reach over 60 miles high and 100 miles distance, providing coverage for relatively large areas. Then, the lower-tier missiles would be deployed to intercept

---

[11] In addition to limiting exports of entire missiles, the MTCR limits the sale of certain components; and here it appears the regime has had some effect in slowing the missile development programs in Argentina and Iran.

[12] See Testimony of Lt. Gen. M. O'Neill, Director of the BMDO, before the Senate Appropriations Subcommittee on Defense, May 24, 1994, pp. 10-14.

those re-entry vehicles that had gotten through the first tier and those that had trajectories under the flight paths of the upper tier. Both sets of interceptor missiles would have use of space-based sensors and ground-based radars, and would have a modern command, control, and communications links. For both the upper and lower-tiers, the BMDO is planning on purchasing systems that can be deployed in the near-term future, while simultaneously working on improved missiles that might be deployed after the year 2000. Table 6.3 below shows the systems under consideration.

It is extremely difficult for an outsider, not privy to classified performance data, to have a fully informed opinion about whether the BMDO is pursuing the appropriate technical directions for its future TMD systems. The basic logic of: (a) having two tiers of defense, and (b) dividing the program into systems produceable now versus those available more than a decade hence appears sound. It is also sensible to have competing programs so that the success of the entire effort does not depend on any one specific system. Yet, whether the particular technologies being proposed for these systems are optimal, will probably only be known as development and testing proceeds.

TABLE 6.3   Present and Potential Future TMD Systems

	Lower Tier	Upper Tier
Present & Near-Term Future	- AEGIS SM-2 - Patriot PAC-3 - ERINT	- THAAD
Longer-Term Future	- Corps SAM - Upgraded Hawk	- AEGIS SM-4 - BPI

Note: Definitions are as follows:
  - AEGIS systems are sea-based to go on surface ships, including the spy 1 radar and the SM series of missiles,
  - ERINT is Extended Range Interceptor,
  - THAAD is Theater High Altitude Defense,
  - Corps SAM is a surface to air missile under design for limited areas such as a location of a corps,
  - Upgraded Hawk is further improvements on the Hawk, currently an anti-aircraft system, and
  - BPI is Boost Phase Interceptor.

Two immediate issues will illustrate the problem of judging which technology to use. A key concern during any war would be to try to find the opponent's missiles and destroy them on the ground. Unfortunately, if the Gulf war experience is indicative, it is extremely unlikely that any substantial percentage of an opponent's missiles will be knocked out in this fashion. In fact, although the U.S. completely dominated the air space after the second day of the war, an Air Force study has acknowledged that there is no convincing evidence that any *mobile* Scud launchers in the field were actually destroyed.[13]

Those favoring the use of deep-strike aircraft would urge greater surveillance capability linked directly to stand-off air strikes or possibly more stealthy ground attack planes. Advocates of TMD would also favor greater surveillance, but would probably say that destroying the launchers on the ground is infeasible. TMD supporters would urge deployment of various kinds of stand-off platforms (aircraft or drones) equipped with interceptors capable of stopping the missiles in their Boost Phase (before decoys could be deployed). Since deep-strike aircraft can be used for a variety of missions and will certainly be maintained in the Air Force inventory, it may take another war and further battlefield experience before this issue can be resolved more definitively.

A second, critical technical question is whether TMD interceptors should be designed to collide directly with an incoming RV (i.e. "hit to kill")? or Is it preferable to have a "proximity fuze" that explodes a charge near the RV? The current Patriot PAC-2 uses a proximity fuze. As mentioned in Chapter 3, some analysts have argued that the fuze was the cause of many misses in the Gulf War, because the incoming Scuds were disintegrating and the Patriot charges exploded at the first debris rather than homing in on the actual RV.

A related claim is that fast-moving, incoming RVs outrun an explosive charge unless the interceptor is directly below the descending RV. The current, lower-tier interceptor favored by the BMDO[14] is the ERINT which is designed as a direct collision or "hit to kill" system. This, obviously, requires

---

[13] For an overview of the air campaign, see, E. Cohen, "The Mystique of Air Power," *Foreign Affairs*, January/February 1994, Vol. 73, No. 1, pp. 109-124.

[14] It should be pointed out that no SDI funds went into the development of the Patriot which was an Army program from the start. Congress has now required that work on TMD systems be managed by the BMDO and integrated with NMD research.

greater precision in the targeting and positioning of the interceptor.[15] Although deciding between proximity fusing and direct impact systems can be done by simulation, there may be some merit in continuing to develop both technologies as they each have important positive features.

Overall, however, in a book with only unclassified information available to us, it is simply not feasible to make a comprehensive judgment about whether the BMDO is taking the right technical directions. The best solution is probably to continue the practice of having outside, independent review boards of technically-qualified individuals cross-checking BMDO recommendations.

Will upgraded TMD systems blur the distinction between "theater" and "strategic" missile defenses?

As discussed above, the U.S. has several programs underway that should produce TMDs which are far more capable than Patriot PAC-2 and PAC-3. The one that is closest to deployment is THAAD which will be ready for testing in 1995. However, the Navy's AEGIS SM 2 Block IV is well underway and the Corps SAM and Boost Phase Interceptor (BPI) could each have functioning prototypes within the next 5-8 years.

Although strong majorities in the Congress and many nations allied with the U.S. are strongly in favor of higher-performance TMDs, four principal arguments have been used against them:[16]

- Improved TMDs will be able to intercept some strategic missiles; and, because they are mobile and could be widely deployed, would be a means for undercutting the ABM Treaty.
- Higher performance TMDs will be a stimulus to the arms race because countries with only moderately sophisticated missile systems will want to upgrade their offensive capability to be able to overwhelm the new TMDs.[17]

---

[15] ERINT has a K-band radar seeker to distinguish decoys from RVs, but also needs exceptional accuracy to maneuver adequately to achieve direct impacts.

[16] For a full elaboration of these points, see, D. Wright and L. Gronlund, "The Security Implications of the Proposed Modifications of the ABM Treaty," *Mimeo.*, (Cambridge, Mass.: Union of Concerned Scientists, December 16, 1994), 9 pp.

[17] For example, France might want to upgrade its missile force if Russia gets better TMDs; similarly, China might want to upgrade its missiles if it perceives the U.S. TMDs as a threat to a Chinese second strike capability.

- The anticipated TMDs will still be vulnerable to certain types of counter-measures (such as cooling of the warheads to foil thermal sensors), so deploying TMDs could stimulate opponents without giving a high level of protection.
- The U.S. could develop more limited TMDs which would comply with the ABM Treaty; and, then through the arms control process, put pressure on other countries not to (a) upgrade their offenses, or (b) develop more sophisticated TMDs of their own.

The Clinton Administration has asserted that THAAD will not be a threat to the ABM Treaty because: (i) THAAD's sensors would not be able to detect incoming ICBM RVs in time given the speed of strategic missiles, (ii) as strategic warheads would be cooled by the long trip in space, they would be too cold for THAAD's sensors to pick up, (iii) THAAD's direct thrusting/ guidance system cannot deal with the high closing speeds necessary to intercept strategic warheads, and (iv) THAAD's computer software is not designed to deal with the short decision-making time that would be necessary to have confidence in intercepting strategic re-entry vehicles.[18]

The Clinton Administration has, nevertheless, been concerned about preserving the ABM Treaty and has proposed to the Russians that the Standing Consultative Committee agree on several "clarifications" of the Treaty. Although the ABM Treaty does not limit TMDs at all and its terms only apply to strategic missiles and systems, there is a flaw in the Treaty because it does not include a definition of the distinction between theater and strategic missiles. This has been a source of debate since 1972 because the SALT I and II agreements defined "strategic" missiles as one with a range longer than 5,500 km. However, exceptions were made to cover, under SALT restrictions, three Soviet missiles (the SS-N-5, the SS-4, and SS-5) that had ranges of less than 5,500 km.[19]

In today's world, where theater missiles are attaining higher speeds (and therefore greater ranges), it is common to define TBMs as those missiles that re-enter the atmosphere at up to 5 km per second and have ranges of up to

---

[18] See the Testimony of Lt. Gen. Malcolm O'Neill before the Military Acquisition and Research and Technology Subcommittees of the House Armed Services Committee, Washington, D.C., June 10, 1993.

[19] See, S. Hildreth, "The ABM Treaty and Theater Missile Defense: Proposed Change and Potential Implications," *Mimeo.*, (Washington, D.C.: Congressional Research Service, May 2, 1994), pp. 2-5.

3,000 km.[20] The Chinese CSS-2, which has been sold to Saudi Arabia, has a range of approximately 3,000 km and is one of the longest range, operational TBMs.

There appear to be three principal areas of the TMD program where overlap with strategic systems might occur: (1) TMD Ground-Based Interceptors (GBIs)[21] may prove sufficiently accurate and versatile to intercept slower ICBMs reentering the atmosphere at speeds which TBMs now achieve; (2) the AEGIS upper-tier TMD interceptors are aboard ship and are therefore mobile and might violate the ABM Treaty's prohibition against mobile ABMs; and (3) the Boost Phase Interceptor, if deployed in areas near Russian ICBM fields, might be able to destroy ICBMs during their slow-moving, lift-off stage.

It should be pointed out that each of these three hypotheticals refers to potential future capabilities, and it may well be that the Russians proceed with their own successful TMDs and are not particularly concerned about this problem. Nevertheless, it is worth addressing these issues, so that the reader can make a judgment about their seriousness.

The *Reentry Speed* issue was noted in Chapter 3. As stated, in 1972, John Foster testified on his interpretation of the distinction between "tactical" and "strategic" ballistic missiles. Foster suggested a boundary for testing TMD systems that would not violate the ABM Treaty.

Sidney Graybeal summarizes these events as follows:

> Although Foster's testimony was never made public, in an unclassified letter to Senator Proxmire, he provided parameters of 2 km/sec velocity and 40 km altitude. He also noted that strategic ballistic missiles reached altitudes well above the atmosphere and peak velocities of 4 to 7 km/sec. The Foster Box was internal Department of Defense guidance for U.S. tests requiring compliance review; it was not an agreed upon dividing line between strategic and tactical missiles, and it was never officially discussed with the Soviets.[22]

An additional complication is that Article VI of the ABM Treaty forbids either the U.S. or Russia from giving non-ABM systems or components the

---

[20] Obviously, for countries that are in close proximity to each other, it makes little difference if they are threatened with a "strategic" or "tactical" missile.

[21] At present, this might apply to THAAD but it would almost certainly apply to more capable TMDs in the future.

[22] Graybeal, S., "Testimony before the Senate Foreign Relations Committee on Effective Theater Missile Defenses and the ABM Treaty," *Mimeo.*, May 3, 1994.

"capabilities to counter strategic ballistic missiles." The Clinton Administration is aware of this problem and has been discussing the issue with the Russians in the Security Consultative Commission (SCC) which was established by the Treaty as the means for resolving this kind of technical issue.

Although the SCC deliberations are secret, it is known that the Clinton Administration position is to "clarify" the ABM Treaty by permitting TMDs capable of handling warheads entering the atmosphere at up to 5 km/sec. Since most modern ICBMs[23] have maximum velocities of over 7 km/sec., this would create a clear difference in performance which both Russia and the U.S. could verify by "national technical means."[24] So, if the Clinton Administration position is accepted by the Russians, TMDs will be able to handle RVs of up to 5 km/sec., then there will be a gap of 2 km/sec., and all RVs reentering at over 7 km/sec. will be considered "strategic."

Opponents of high-performance TMDs argue that TMDs capable of handling RVs of up to 5 km per second will also be able to intercept a sufficient number of ICBMs to constitute quasi-effective ABM systems. For example, the Union of Concerned Scientists (UCS) has done simulations estimating that a TMD system like THAAD could defend a region the size of a small metropolitan area from a limited ICBM attack.[25] The UCS thus concludes that improved TMDs will undercut the ABM Treaty and encourage opponents to upgrade their offensive systems to be able to overwhelm TMDs. Hence, some opponents of upgraded TMDs now favor limiting the speed and maneuverability of TMDs to ensure that strategic missiles cannot be intercepted.[26]

This controversy has caused a rift between the American arms control community and the Clinton Administration[27] which most supporters of missile

---

[23] The Russian submarine launched ballistic missile, the SS-N-6 has an intermediate range (3000 km) and reenters the atmosphere at about 5 km/sec., so it could potentially be stopped by the newer TMDs. However, the Russians have retired most of the Yankee class submarines that use the SS-N-6, and this should not be a major problem to negotiate.

[24] "National Technical Means" is the euphemism used in the ABM Treaty for various types of radar and thermal monitoring techniques used by both the U.S. and Russia to track each other's tests.

[25] Gronlund, L., G. Lewis, T. Postol, and D. Wright, "Highly Capable Theater Missile Defenses and the ABM Treaty," *Arms Control Today*, April 1994, pp. 5-8.

[26] Wright, D. and L. Gronlund, "The Security Implications of the Proposed Modifications of the ABM Treaty," *Mimeo.*, (Cambridge, Mass.: Union of Concerned Scientists, December 16, 1994), p. 8.

[27] See, for example, the exchange between the Clinton Administration's Director of the Arms Control and Disarmament Agency, John Holum, and John Pike of the Federation of American Scientists. J. Holum, "Don't Put the Allies at Risk," and

defense see as an unnecessary diversion from the goal of attaining improved TMDs. Nevertheless, if the Russian Government will not agree to the "clarifications" of the ABM Treaty, the U.S. will have to decide whether just to proceed unilaterally and certify that THAAD is "treaty compliant" or whether to slow progress for TMD upgrading.

The *AEGIS SM-4* poses a variant of the missile-speed reentry question. If the current AEGIS SM-2 missile is upgraded to handle RVs coming in at higher altitudes and speeds,[28] then it too, like the Ground Based Interceptors, could conceivably block slower-moving ICBMs. The SM-2s and SM-4s are designed to give a naval task force protection against missile attack while it is deployed overseas.[29] Yet, the further complication is that ship-based TMDs could clearly be moved aboard ship back to any of dozens of ports around the U.S. (providing North America with a large number of quasi-ABM sites).

If AEGIS SM-4 were really capable of intercepting ICBMs, then it would be a "mobile" ABM which is forbidden by the 1972 Treaty; and, if moved into U.S. ports, it would provide more than the one National Missile Defense site agreed in the 1974 protocol to the Treaty. Since the AEGIS SM-4 has not yet been developed and tested, this is not an immediate issue, but one that needs to be addressed as long as the U.S. intends no major revisions of the ABM Treaty. The most direct way to resolve this problem would be to have performance levels for AEGIS SM-4 roughly equivalent to the THAAD ground based interceptor, so they would be doing the same function in different locations.

The *Boost Phase Interceptor* (BPI) issue is even more complicated. Designers of TMD systems are now facing the same problems that NMD engineers have worried about for three decades. Formerly, TBMs traveled relatively short distances (50 to 500 km) and had trajectories that never or only briefly went outside the atmosphere. This meant that it was not generally feasible to use decoys with TBMs.

At present, however, TBMs are traveling higher and spending more time outside the atmosphere. This gives the attacker the chance to: (a) use

---

(Continued)

J. Pike, "Don't Imperil the Treaty," *New York Times*, October 25, 1994, p. A-21.

[28] The advantage of intercepting RVs at higher altitudes is that one interceptor can thus defend a larger area. However, at high altitudes, the RVs (and possibly decoys) have not yet been slowed appreciably by the atmosphere, so successful "upper-tier" TMDs have to be able to intercept at higher speeds.

[29] The coverage anticipated for upgraded naval TMDs would defend a task force at sea and, additionally, protect staging areas on the ground if an amphibious landing were necessary.

decoys, or (b) in the case of chemical or biological warheads, disperse a large number of "bomblets" that could be used to target a large grid area.[30] Thus, there may be a real premium placed on intercepting and destroying sophisticated TBMs while they are still in their boost phase, before they can disperse decoys or bomblets. Hence, boost phase intercept looks as if it is the optimal TMD (much as it does for ICBMs).

The difficulty is that TMD Boost Phase Interceptors deployed within range of ICBM silos could do double duty: potentially destroying either TBMs or ICBMs. This would be expensive because the BPIs would need to be deployed on aircraft or drones that were in the air at all times. Also, at currently discussed ranges for the BPIs, the only ICBM fields within range of U.S. aircraft on the perimeter of Russia and the Commonwealth of Independent States would be the sites in Kazakhstan.

Nevertheless, there is no inherent reason why BPI ranges could not be extended, so the Russians would certainly see them as potential mobile ABMs. Like the naval TMDs, the BPI is only in the early stages of development, so this does not need to be resolved now. Moreover, in the current cooperative strategic environment, the BPI issue might be handled by negotiating limits on the ranges for the interceptor and locations where they will be deployed.

In the long-run, however, future BPIs may prove sufficiently effective that they tilt the balance in favor of missile defense over offense. This could lead to a proliferation of various sorts of BPIs, and might even force the major powers to shift away from reliance on offense and Deterrence to a defense-based strategy.

Should the U.S. Have a National Missile Defense System?

The answer the Clinton Administration has given to this question is: not now, but we should continue work on a technology program that gives the U.S. the ability to deploy an NMD within five years if the strategic picture changes and it appears to be warranted. Therefore, unlike the Bush Administration, which supported the deployment of Ground Based Interceptors as soon as feasible and Brilliant Pebbles after testing, the Perry Defense Department is taking a much lower-keyed approach to NMD.

As noted in Chapters 4 and 5, the U.S. could deploy one site of GBIs with

---

[30] There is also the possibility that opponents could cool their warheads or cover them with radar-deflecting materials, thus limiting the effectiveness of thermal sensors and radars.

up to 100 launchers and 100 interceptors and still be in full compliance with the 1974 Protocol of the ABM Treaty.[31] There is no limit on the range of the GBIs, so they could, potentially, be designed to intercept incoming RVs beyond U.S. territory. The Treaty contains restrictions on the location of large-scale phased array radars (LPARs)[32] but no limits on sensors as long as the sensors do not communicate directly with interceptors. Most of the Treaty's technical restrictions were included on the assumption that they preserved "strategic stability" by limiting the effectiveness of missile defense. This was done to build confidence that both sides would have assured retaliatory capability and no incentive to strike first in a crisis situation.

One of the fundamental problems with the ABM Treaty as a framework for U.S. defense in the 1990s and beyond is its premise that vulnerability is stabilizing. That may well have been true for the U.S. and Soviet Union when they had rough parity in nuclear weapons, but it is probably not the case when the U.S. is dealing with small nuclear powers and very asymmetrical force balances.

There is no third-world country[33] currently capable of reaching U.S. soil with its missiles. Moreover, the first new nations that do achieve that status will almost certainly have small inventories. These would not be plausible weapons for a comprehensive attack. However, if the U.S. continues to have no National Missile Defense, a country with even a small inventory of ICBMs could threaten an attack or retaliate if its own interests were at stake.

There have been two major studies of the future ICBM threat to the U.S. The most highly publicized assessment was by the CIA. In testimony, Director of Central Intelligence James Woolsey said:

> Ballistic missiles are becoming the weapon of choice for nations otherwise unable to strike their enemies at long ranges. Today there are 25 countries, many hostile to our interests, that are developing

---

[31] The GBIs must be within a radius of 150 km and centered on either a national capital or a missile field. Rapid reload launchers are banned, and Agreed Statement E forbids interceptors with multiple independently targetable reentry vehicles (MIRVs).

[32] LPARs must be placed on the edges of a country's borders and must face outward - to make them capable of identifying an attack but not part of a battle management system that directly cues or redirects interceptors in flight.

[33] China, which now has the world's third largest GNP and is a middle power in military terms, is the one possible exception to this rule. It has a small number of ICBMs, but they are hidden in remote places and clearly designed as retaliatory weapons, not as ones ready for a first strike.

nuclear, chemical, and biological weapons..... Moreover, some of these countries may place little stock in the theory of deterrence....Biological weapons are a particular concern, especially given the ease of setting up a laboratory and the difficulty in distinguishing between dual-use products. It's hard to get international consensus to condemn a supplier or user of such dual-use materials or technology.[34]

Lt. Gen. James Clapper, Jr. of the Defense Intelligence Agency, testified at the same hearing and commented:

> While no additional countries will be able to reach the United States with missiles this decade, proliferation of missiles and weapons of mass destruction are of particular concern and a very tough problem for intelligence.... Countries who want the technology bad enough will eventually get it, and those that participate in these activities are becoming increasingly resourceful at masking such activities.[35]

The most comprehensive, non-governmental look at this subject was done in 1993 by the Proliferation Study Team, chaired by Gen. William Odom, USA, (ret.). The Study Team concurred with the CIA view, that the probability of new, long-range ballistic missile threats to the U.S. developing during the 1990s was quite low. Yet, their report stressed four ways in which hostile countries might accelerate their plans and pose an earlier missile threat to the continental U.S.:

1) acquiring ballistic missile components from sources in China, Ukraine, or Russia (possibly without approval by the governments in Beijing, Kiev, or Moscow);
2) indigenous manufacture or purchase from a foreign supplier of space launch vehicles that could be converted to military use as ICBMs;
3) a rapid decline in the security environment, leading countries that previously had the technical and economic capability to develop offensive missiles but did not deploy them, to reverse position and establish ICBM programs;[36] and

---

[34] Woolsey, J., "World Trouble Spots," Hearing Before the Senate Select Intelligence Committee, January 25, 1994, (Washington, D.C.: Federal News Service), p. 9.

[35] Clapper, Lt. Gen. J., Jr., "Remarks" at the Senate Select Intelligence Committee Hearing, January 25, 1994, (Washington, D.C.: Federal News Service), p. 16.

[36] China, Great Britain, France, and Russia today all possess the capability to

4) the purchase of shorter-range missiles by countries in Latin America or the Caribbean or the use of their territories by hostile third parties.[37]

Obviously, it is not possible to predict any of these potential developments with precision, but the key point of the Proliferation Study Team's report was to emphasize that change could come more quickly than would be anticipated by looking at the technical capacities of potentially hostile powers.

The Proliferation Study Team also raised another key issue. It noted that major international realignments can take place within a 10-20 year period, and countries that were previously close can become antagonists and opponents can become allies.[38] Hence, the Team concluded it is only prudent to anticipate that some country will either move faster in acquiring ICBMs than currently predicted or that alliance restructuring may produce an opponent not presently expected.

Although the Proliferation Study Team's methodology was different than that being used in this volume, the members came to essentially the same conclusion: National Missile Defense is designed to deal with low probability threats, but they are definitely not zero probability threats and the U.S. must be prepared to respond on short notice.

At present, there are three principal schools of thought about whether to deploy National Missile Defense[39] for the continental U.S. Their positions can be summarized as:

1) NMD would be a grievous mistake.
2) NMD is not needed now, but the U.S. should retain the option of deploying it within 5 years if the threats become tangible.
3) NMD will be needed by the year 2000, so begin procurement and deployment planning now.

(Continued)

reach the U.S. with ballistic missiles. Depending upon circumstances, Belarus, Ukraine, Kazakhstan, Brazil, India, Italy, Israel, Germany, Sweden, and Japan could all, conceivably, develop that capacity in the 1990s.

[37] Odom, Lt. Gen. W.E., USA (ret.), Chairman, The Proliferation Study Team, *The Emerging Ballistic Missile Threat to the United States*, (Washington, D.C.: February 1993), p. 1.

[38] The three cases cited were: (a) the Franco-British rapprochement at the end of the 19th Century, (b) the development of Anglo-American cooperation at the beginning of the 20th Century, and (c) the Sino-Soviet Split in the 1960s.

[39] At this time, there is no sizable group advocating the immediate deployment of Brilliant Pebbles, so, when NMD is referred to here, it will be assumed to be only a Ground Based Interceptor system.

The committed *opponents* of NMD make many of the same arguments against it that they use against upgrading Theater Missile Defenses: (a) the threat is uncertain and nations will certainly be deterred by even the post-START II level of U.S. nuclear inventory and offensive capability; (b) building an NMD could prejudice chances for future arms reduction efforts; and (c) the 100 Ground-Based Interceptors permitted under the ABM Treaty are, by design, too few to provide a comprehensive defense, but deploying them would create a momentum to undercut the entire Deterrence regime.

Each of the arguments of the committed opponents can be successfully rebutted. True, rational governments would be impressed by the U.S. retaliatory capability, but Saddam Hussein was not willing to withdraw from Kuwait even after being given a clear deadline when he could see the massive forces arrayed against him.

It is inconceivable that Iraq would have been more conciliatory had it possessed ICBMs with nuclear weapons, and it is very possible that the U.S. would have been unable to mount opposition to Saddam Hussein if the force balance was even slightly more favorable to him.[40] In the future, countries willing to accept substantial casualties would then gain an enormous advantage if they also got functioning ICBMs. Thus, the limits of Deterrence are clear, and "assured second strikes" may not deter future hostile nations as they did with the Soviet Union.

The argument that building a National Missile Defense will prejudice future arms control negotiations is curious, particularly because the Soviet Union / Russia has had a functioning NMD since before the signing of the ABM Treaty. The more interesting question is whether deploying 100 GBIs would create enough momentum for missile defense that it would undercut the support for Deterrence. Obviously, advocates for missile defense hope that a small NMD deployment would legitimize the concept, while detractors oppose even having capabilities allowed under the ABM Treaty.

Clearly the real maneuvering here has to do with whether the ABM Treaty will eventually be renegotiated or terminated. If 100 GBIs were deployed in the U.S. and they either worked successfully during an attack or appeared to

---

[40] For example, many prominent defense experts, including Former Secretary of Defense, James Schlesinger, and Former Chairman of the Joint Chiefs, Adm. William Crowe, argued against committing U.S. forces in the Gulf War on the grounds that casualties would be high and that the continuation of sanctions might work. If Iraq had more threatening weapons, the chorus of opposition to the Persian Gulf War would have been even greater.

deter during a future crisis, the public and Congressional support for more missile defense would overwhelm the arms control community's opposition.

It is also possible that Russia (which is far closer than the U.S. to potential trouble spots in the Middle East, South Asia, and the Korean peninsula) may want to renegotiate the ABM Treaty.[41] Thus, we have a situation where opponents see any GBI deployments as the initial step toward a flawed change in U.S. security policy, whereas many advocates favor a thin NMD as: (a) desirable in itself, and (b) the first step in a transition away from pure Deterrence.

The position of those urging *no decision on deployment, concentrate on Research and Development* is usually based on the assumption that there will be no new ICBM threat to the U.S. in the 1990s. This is the view that former Secretary of Defense Aspin took in the "Bottom-Up Review" which was apparently endorsed by President Clinton as well as Aspin's successor, Secretary William Perry. One principal appeal of this position is that it saves money in the short-run. Mostly by cutting back on NMD, the Clinton Administration proposed to trim back the combined NMD and TMD programs from $39 to $17 billion in Fiscal Years 1995-1999.[42]

Putting off deployment and concentrating on R & D has another advantage: it avoids committing to a particular mix of current technologies and allows for the possibility of improved performance in components or systems that may become available later in the decade or after the year 2000.

The obvious disadvantage of waiting, however, is that the U.S. will not have a functioning NMD in this century. Moreover, it would take about 5-8 years after a decision is made to deploy before a functioning NMD could be in place. Thus, the central issue is: will the cost savings be worth the loss of protection inherent in foregoing the NMD?

Advocates of *deployment of NMD before the year 2000* include differing viewpoints,[43] but there is common agreement on several reasons for proceeding now with an initial GBI system:

---

[41]   As noted in Chapter 5, it was President Yeltsin who in 1992 openly advocated a Global Protection System for missile defense. Yeltsin dropped that initiative, however, when it was clear that President Bush (who supported GPALS) would lose the November election and there was opposition in the Russian military.

[42]   Aspin, L., "The Bottom-Up Review," *Mimeo.*, (Washington, D.C.: U.S. Department of Defense, September 1, 1993), p. 23. The original $18 billion figure was, almost immediately, cut to $17 billion; and, after the August 1994 revisions, was cut to $15.2 billion for the five years up to FY 1999.

[43]   It is reasonable to divide them into three groups: (1) those urging only a GBI system, fully compliant with the narrow interpretation of the 1972 ABM Treaty,

(i) the 5-8 year lead time for manufacturing and testing of such a complex system means that, if the U.S. wants this type of insurance for early in the next century, it must decide now;

(ii) deploying and managing operational systems is different from running R & D programs, and, if the U.S. wants to develop a group of individuals competent to handle missile defense systems, they need to have operational batteries to work with;

(iii) evaluating the pros and cons of functioning systems is a key aspect of making design improvements for the next generation of missile defense; and

(iv) the decision to deploy sends an important signal that the U.S. will not leave itself vulnerable to small powers, and the rest of the world should begin the transition process of adjusting strategy from exclusive reliance on offensive systems to a new mix of offensive and defensive ones.

The view, in this volume, is that deployment of a GBI NMD is a desirable goal and should begin in the next few years. If there was a compelling technical reason to delay for testing of an essential element of the system, a loss of a year or two in deployment would not appear to be critical. Yet, to forego NMD indefinitely (as some in the arms control community urge) or to keep NMD as only an R & D program risks becoming vulnerable to the evolving ICBM threat in the early part of the next decade.

If the U.S. should have NMD, what should its position be on the ABM Treaty? There are at least three aspects of the "treaty compliance" problem that need to be considered: (a) reactions by Russia, (b) reactions by third party governments, and (c) domestic political concerns inside the U.S.

Unless Russian President Yeltsin has fundamentally changed his view since 1992[44] the Russian Government should not be concerned if the U.S. pursues a Treaty-compliant deployment of 100 GBIs.[45] The more difficult

---

(Continued)

(2) those favoring a GBI system but with up to 5 sites, requiring a modest renegotiation of the ABM Treaty, and (3) those who would deploy GBIs but favor Brilliant Pebbles as well, who would be willing to fundamentally renegotiate or withdraw from the ABM Treaty.

[44] Yeltsin's reasons for putting off serious negotiation on the Global Protection System, as conveyed by his envoy Gen. Mamadoff, were tactical and reflected the politics of 1992 - not a reversal of view about the desirability of missile defense.

[45] For cogently argued support of deploying a Treaty-compliant GBI system, see, S. Graybeal and P. McFate, *The ABM Treaty and Ballistic Missile Defense: Can the Circle Be Squared?*, (Washington, D.C.: American Association for the

issue is whether the U.S. should now broach the question of amending the Treaty to permit more than one site. President Yeltsin's militant and nationalistic recent moves, particularly in dealing with the Chechen secession crisis, are worrying however and might presage more tension in U.S.-Russia relations.

Although putting the GBIs at Grand Forks, N.D. would be an acceptable choice for a thin defense of the continental U.S., it would not be adequate for threats against Hawaii and Alaska. Also, one site in the central part of the U.S. is not very effective at countering "accidental launch" or firings by a "rogue commander" from a submarine. Hence, there is some merit in spreading the GBIs at four to five U.S. locations.

Russia's size would make multiple GBI sites attractive to them as well, but the exceptional economic difficulties the country is facing make it unlikely that Russia would want to proceed with those military expenditures now. Therefore, it is quite possible the Russians would demur on renegotiating the Treaty because they would not want the U.S. to proceed with a modern NMD system, while they stayed with their outdated Galosh interceptors.

Since the Russians have already deflected serious discussions on Brilliant Pebbles deployments, the U.S. might find that the current circumstances are simply not fortuitous for anything more than the 100 GBIs at one site, but that could certainly proceed apace without the concurrence of other nations.

This leads to the issue of third party reactions. The Clinton Administration has taken the position that all interpretations and potential revisions of the ABM Treaty should be negotiated in a multilateral context, not just directly with the Russians. This is a most unfortunate, self-imposed stance which will vastly complicate all clarifications and modifications of the Treaty. This position was taken with the intent of placating Ukraine and Kazakhstan, but it is likely to delay, needlessly, all bargaining that relates to the Treaty.[46] There is also the question of how the Chinese will see these negotiations if they are not included. The Clinton decision to sell a variant of Patriot to Taiwan[47] has

---

(Continued)

Advancement of Sciences, Program on Science and International Security Occasional Paper, No. 93-26S, 1993).

[46] In addition, the Clinton position gives Ukraine and Kazakhstan an incentive to cooperate with Russia (rather than the U.S.) so that negotiations can be blocked and they can bargain for concessions from the U.S. on (economic) issues that really matter to them.

[47] Taiwan Minister of National Defense, Sun Chen, announced that the first payment will be for $170 million, with initial deliveries beginning in 1996. See, *Far Eastern Economic Review*, July 14, 1994, p. 13.

already attracted Beijing's attention to these issues. Thus, it would be sensible for Washington to quickly restate its position, and handle these discussions on a bilateral basis, limiting the agenda to only those topics that are specifically relevant to the country involved.

The question of domestic political support for NMD is probably the major stumbling block. Although the Congress did pass the Missile Defense Act of 1991 authorizing planning for and purchasing of a GBI system by 1996, ever since that time, support for NMD has waned. The Clinton Administration favors only an NMD "technology readiness program," but the recently-elected Republican majorities in the House and Senate may, ultimately, accelerate the NMD effort.

If the U.S. Proceeds with Limited Ballistic Missile Defense, What Related Changes in Policy Should Occur?

As this book goes to press, the most volatile and potentially dangerous foreign policy problem the U.S. faces is: what to do if North Korea backs down on its October 1994 commitments to stop its nuclear program and begin adherence to a phased inspection plan? The dilemma has been heightened by North Korea's conventional arms build-up, including the purchase of submarines[48] and the development of its own cruise missiles.[49] All of these factors have been further complicated by the death of Kim Il Sung, the country's leader for 49 years and his son's uncertain hold on power.

The complexity of the North Korean problem is extraordinary, not only because of the number of countries concerned but because so much of the information that outsiders would like to know is ambiguous or being consciously manipulated by the government in Pyongyang. It is now widely assumed that North Korea has already produced one or two nuclear weapons, and that the rods removed from the Yongbyon reactor in mid-April 1994 could give them the potential to make four or five additional nuclear bombs.[50] Despite assurances that the rods will eventually be inspected, one of the least satisfactory aspects of the U.S.-North Korean agreement is that the precise disposition of these rods is left vague.

---

[48]   Sanger, D., "North Korea Buying Old Russian Subs," *New York Times*, January 21, 1994, p. A-6.

[49]   Gordon, M., "North Korea Test Cruise Missile Designed to Sink Ships," *New York Times*, June 1, 1994, p. A-12.

[50]   Gordon, M., "North Korea Said to Have A-Bomb Fuel - Pace of Program Alarms Washington," *New York Times*, June 8, 1994, p. A-7.

Although there is no evidence that North Korea has tested a nuclear weapon, its delivery systems are formidable (aircraft and the Rodong I and II missiles), so Pyongyang has already built a shadowy deterrent. Even if this particular situation is resolved without conflict, we can assume that other countries will learn from the North Korean example. Others are likely to acquire or openly develop delivery systems and secretly work on nuclear and chemical / biological weapons.

In addition to proceeding with upgrading TMDs and deploying an NMD so that potentially hostile countries know we will not be coerced, the U.S. needs to consider two additional steps:

1) providing a clearer set of incentives and disincentives for countries that cooperate with or attempt to thwart non-proliferation of weapons of mass destruction, and

2) recognizing that the demand for BMD systems will grow and that the U.S. should distinguish between non-threatening defensive systems and offensive systems.[51]

As the twin issues of nuclear capability and missile capability become increasingly troublesome, the U.S. may want to consider using a mix of policies that tangibly encourages cooperative behavior. For example, in thinking about groups of countries, it may be worthwhile to categorize them as: allies, competitors, and hostile states. Allies are those (like Germany and Great Britain) that cooperate actively on conventional force and arms control matters; competitors are those (like India and China) who have divergent strategic interests but are not directly challenging; and hostile states are ones like Iran, Iraq, or Libya that take active steps to adversely affect the U.S. Table 6.4 below is a highly simplified schematic, but is designed to show how different areas of policy might be coordinated to deal with the weapons proliferation dilemma.

It is often argued that *economic openness* encourages cooperative behavior among states because the more modern and outward-looking groups are rewarded through trade and foreign investment projects. This may be accurate in many cases and fits nicely with preferred values in the U.S. However, the risk of the spread of weapons of mass destruction (chemical, biolog-

---

[51] In certain cases there is potential overlap between offensive and defensive systems. As discussed, space launch vehicles can be converted to ICBMs, but most defensive missiles have neither the payload nor range to be effective offensive weapons.

TABLE 6.4 Policy Changes to Supplement BMD

Policy Arenas	Toward Allies	Toward Competitors	Toward Hostile Nations
Political & Economic Measures	Encourage open economies	Use positive or negative suasion	Isolate diplomatically
Arms Control	Permit BMD sales	Discourage offensive weapons	Limit all sales
Industrial Policy	Expect real reciprocity	Further modify COCOM	Limit all key technology flows
Military	Support alliances	Encourage defensive Strategies	Discourage their alliances

ical, and nuclear) is sufficiently threatening that the U.S. should be willing to think of all feasible means for encouraging a stable international environment. The U.S. has both the legislation for and a history of using economic sanctions for openly hostile states, but we have not yet thought out how
best to reward or discourage those nations that are clearly pursuing actions adverse to our interests. It may well be that we need to use access to or denial from the U.S. market in a more directly dirigiste fashion, so competitor nations get clear signals and suffer consequences if they continue with counter-productive policies..

U.S. *arms control* policy is in turmoil today precisely because President Clinton set very specific standards for North Korean compliance with the

Non-Proliferation Treaty (NPT),[52] but was not willing to back up those claims with force and found that the neighboring countries (China, South Korea, and Japan) were unwilling to use even strong economic sanctions.

One clear lesson from this episode is that major powers should not make claims they are unwilling to enforce; but the more fundamental issue is: how to keep new nations from getting as far in the nuclear development process as North Korea achieved by 1994? There is, obviously, no single answer to this quandary, yet it seems that more effort should be focused on limiting transfers of specific high-tech components rather than arms limitations in general. Moreover, it may be stabilizing to encourage the sale and development of defensive systems. Hence, more effort needs to be put into distinguishing useful versus destabilizing arms transfers.[53]

Industrial policy is a tool which most nations use quite openly, whereas, in the U.S., there is some embarrassment about it. American industrial policy efforts are generally labeled "R & D" or "cooperative programs." This is mentioned here because large amounts of important technology are transferred to allies and friendly nations via "coproduction" and "codevelopment" projects. The U.S. needs to decide if it wants to continue these programs; and, if so, what criteria should be used for ensuring reciprocity from those nations benefiting from our largesse.[54] Also, as the overall global level of technological sophistication grows, there will be a need to focus on limiting those items and materials with the greatest military significance, whether they are available through sales, military cooperative programs, or civilian projects making dual-use products.[55]

Changes in *military policy* as a result of greater use of BMDs would be far-reaching. There is not space here to elaborate these in depth, but the changes would be in related hardware, doctrine and strategy. Using BMDs will probably be a part of a broader set of changes in military thinking and war-fighting methods. In the future, units are likely to be smaller and there

---

[52] The President and his aides repeatedly stated in 1993 and 1994 that it was "unacceptable" for North Korea to have a nuclear weapon.

[53] For a debate on these distinctions, see, S. Gejdenson and T. Roth, "Export Policies for the 1990s," and G. Milholin and G. White, "Proliferation in Disguise," in the *New York Times*, July 18, 1994, p. A-15.

[54] See, D. Denoon, *Real Reciprocity - Balancing U.S. Economic and Security Policies in the Pacific Basin*, (New York: Council on Foreign Relations, 1993), p. 87.

[55] For example, cellular phones and pleasure-boat positioning systems can be immediately transferred to military use.

may be less effort to defend platforms (key aircraft and capital ships) and more emphasis on mobility and maneuver. Also, logistical areas will become more vulnerable to missile attack so they will need increased BMD protection. Because of improved sensors and "real-time identification techniques," there is also likely to be a greater premium placed on information processing and decentralized decision-making.[56]

Making BMDs effective is going to require continual upgrading of the technology involved. For example, there will need to be better links between surveillance systems and BMD command and control facilities, and links between BMDs and multi-purpose platforms like aircraft and drones. However, perhaps the biggest changes will be in the doctrine of actual military engagements and in the strategic choices about which threats are really the most troubling. It may be that small countries which invest significantly in offensive systems will warrant more concern than large, well-armed nations which focus on defensive measures.

This menu of related policy changes is not meant to be exhaustive, but rather indicative of the breadth of adjustment that may be necessary if the U.S. is to take full advantage of its own shift toward BMDs and less reliance on purely offensive capability.

Overall Conclusions from This Book

Each chapter of this volume has been designed to stand on its own, so this is only a brief summary to recapitulate the principal findings:

1) The launching of SDI was an example of "top-down" decision-making and leadership. At least initially, it was not like many other defense programs which are pushed by an "iron triangle" of supporters from Congress, contractors, and the career military.

2) SDI was meant to achieve a dramatic change in U.S. military thinking and force structure, ending reliance on Deterrence and developing a comprehensive defense for the population at large.

3) SDI failed in its basic objectives for three reasons: (a) The technology did not exist to provide a comprehensive missile defense in the 1980s; (b) Because the goals were overstated, it was relatively easy for opponents (the arms control community, many scientists, and many

---

[56] For a summary of these potential changes, see, T. Ricks, "How Wars Will Change Radically Says Pentagon Planner," *Wall Street Journal*, July 14, 1994, p. 1.

in the U.S. Air Force and Navy) to challenge the credibility of SDI's plans; and (c) The 1987 "Early Deployment" effort by advocates of SDI (pushing for a governmental decision to begin procurement before the mid-1990s) hurt the effort still further.

4) Nevertheless, SDI had important positive effects: (a) It convinced the Soviet leadership the U.S. was demonstrably ahead in military technology and that it was not feasible for the USSR to achieve parity if the U.S. shifted from Deterrence to Defense; and (b) By 1991, when Boris Yeltsin became President, he fully recognized the futility of further arms competition and proposed shifting from offensive to defensive systems, urging the establishment of a Global Protection System of BMDs.

5) It was the use of the Patriot TMDs in the Gulf War and the fears of instability in the USSR after the coup attempt against General Secretary Gorbachev that rekindled interest in BMDs. The press and the general public recognized that there was a threat from ballistic missiles which was not adequately addressed by U.S. reliance on the threat to retaliate. Despite continued opposition from the arms control community, Congress passed the Missile Defense Act of 1991.

6) Interest in BMD has waned again in the middle 1990s as the turmoil in the states of the Former Soviet Union has subsided and the desire to cut the defense budget outweighed the commitment to military modernization.

7) The U.S. could benefit today from having an effective, thin National Missile Defense and upgraded Theater Missile Defenses. The TMDs are an immediate need and the thin NMD is a reasonable form of catastrophic insurance.

8) If BMDs are to be deployed, they should meet four criteria: (i) promote strategic stability, (ii) be survivable against direct attack, (iii) provide greater flexibility in U.S. foreign and defense policy, and (iv) be sufficiently effective that the present value of their protection will be greater than the cost of their procurement.

9) If the U.S. does move in the direction of deploying substantial missile defenses, it will need to make a series of related policy changes. Political and economic policies, arms control, industrial policy, and military doctrine and strategy will each need to be rethought to fully capitalize on this major shift in U.S strategy.

Appendix A

List of the 13 Major Agreements Between the US and USSR from 1972 to 1979[1]

1) Treaty Between the United States of America and the Union of Soviet Socialist Republics on the Limitation of Anti-Ballistic Missile Systems (The ABM Treaty)
    (Signed May 26, 1972 by R. Nixon and L. Brezhnev)
    Articles I to XVI.

2) Interim Agreement Between the United States of America and the Union of Soviet Socialist Republics on Certain Measures with Respect to the Limitation of Strategic Offensive Arms (SALT I Accords)
    (Signed May 26, 1972 by R. Nixon and L. Brezhnev)
    Articles I to VIII.

3) To the Interim Agreement Between the United States of America and the Union of Soviet Socialist Republics on Certain Measures with Respect to the Limitation of Strategic Offensive Arms (SALT I Protocol)
    (Signed May 26, 1972 by R. Nixon and L. Brezhnev)

4) SALT I : Agreed Interpretations and Unilateral Statements

    I. <u>Agreed Interpretations</u>

        A) Initialed Statements of the ABM Treaty

---

[1]This appendix is a synopsis with certain key provisions highlighted. For complete texts of all these documents, see C. Blacker and G. Duffy, eds., *International Arms Control - Issues and Agreements*, 2nd ed., (Stanford: Stanford University Press, 1984).

Section A: Retention of non-phased array radars

Section B: "Potential of the accepted phased array radars

Section C: Distance between the capital city ABM and the second ABM shall be at least 1300 km.

Section D: No phased array radars above a "potential" of 3 million.

Section E: Any system based on "other physical principles" will be subject to discussion.

Section F: Interceptor missiles are limited to targeting 1 incoming reentry vehicle each.

Section G: The U.S. and USSR are forbidden to supply plans for or parts of ABM systems to third countries.

B) Common Understandings

a) Increase in ICBM Silo Dimensions

b) Location of ICBM Defenses

c) ABM Test Ranges

d) Mobile Land-Based ABM Test Ranges

e) Standing Consultative Commission

f) Standstill: Act as if agreement is in force pending ratification

II. Unilateral Statements

Made by the US Delegation:

a) Withdrawal from the ABM Treaty

b) Land-Based Mobile ICBM Launchers

c) Covered Facilities

d) "Heavy" ICBM's

e) Tested in ABM Mode

f) No-Transfer Article of ABM Treaty

g) No Increase in Defense of Early Warning Radars

Made by the Soviet Delegation:

a) Regarding US and NATO Submarines

5) Basic Principles of Relations Between the United States of America and the Union of Soviet Socialist Republics

(Signed May 29, 1972 by R. Nixon and L. Brezhnev)

6) Memorandum of Understanding Between the Government of the United States of America and the Government of the Union of Soviet Socialist Republics Regarding the Establishment of a Standing Consultative Commission
(Signed December 21, 1972)

7) Standing Consultative Commission on Arms Limitation: Protocol and Regulations
(Signed May 30, 1973)

8) Treaty Between the United States of America and the Unions of Soviet Socialist Republics on the Limitation of Underground Nuclear Weapon Tests
(Signed July 3, 1974 by R. Nixon and L. Brezhnev)

9) Protocol to the Treaty Between the United States of America and the Union of Soviet Socialist Republics on the Limitation of Underground Nuclear Weapon Tests
(Signed July 3, 1974 by R. Nixon and L. Brezhnev)

10) Protocol to the Treaty Between the United States of America and the Union of Soviet Socialist Republics on the Limitation of Anti-Ballistic Missile Systems
(Signed July 3, 1974 by R. Nixon and L. Brezhnev)

11) Joint Statement on Strategic Offensive Arms Issued at Vladivostok
(Signed November 24, 1974 by G. Ford and L. Brezhnev)

12) Conference on Security and Cooperation in Europe, Final Act
Document on Confidence-building Measures and Certain Aspects of Security and Disarmament

13) Treaty Between the United States of America and the Union of Soviet Socialist Republics on the Limitation of Strategic Offensive Arms
(Signed June 18, 1979 by J. Carter and L. Brezhnev)

Appendix B

Soviet Approaches to BMD: A Chronology[1]

1945-66: Initial BMD Effort

1948   Soviet National Air Defense Troops (PVO) acquired the status of an independent service. Subsequently, in the 1950s, its role and budget were expanded.

Late 1940s   Research work on a BMD program began in the USSR.

1952   Surface-to-air missile technology became available to the Soviet PVO.

Mid-1950s   The Soviet leadership decided on the development of the Moscow BMD system and began to develop their first BMD components.

1955   The Soviet concept of a "preemptive strike" is proposed as a damage-limiting and retaliatory measure if attack is initiated by the enemy.

The Soviet SA-1 SAM deployment around Moscow began. By the end of the 1950s, the total number of SAMs protecting Soviet cities reached 1,000, and the number of Soviet interceptor aircraft grew to 3,700.

1956   The construction of a BMD test site at Sary Shagan (Kazakhstan) began.

A prominent Soviet academician, P. Kapitsa, claimed that the eventual

---

[1]Special thanks to Mr. Oleg Batrachenko and Mr. Li Wei for their extensive help in researching and drafting this Appendix.

deployment of ABMs as well as long range ballistic missiles would pose new challenges to disarmament, and called for including BMD along with nuclear weapons and delivery systems in general disarmament. However, this was Kapitsa's statement and the Soviet disarmament and arms control position on BMD was not developed until later. In the 1950s and up to 1967, the prevailing view was that BMD would enhance strategic stability.

Aug. 1957   The first successful Soviet launch of an ICBM took place, with the launch of "Sputnik" two months later.

Nov. 1957   Soviet Defense Minister Marshal P. Malinovsky recommended that more attention be paid to air defense and BMD. The same year an initial test of a first generation Soviet ABM took place. Thus, the USSR began to develop offensive and defensive strategic weapons in parallel in the early-to-mid 1950s.

Late 1950s   Development of the Hen House early warning radar began concurrently with the early Soviet BMD system. These radars were tested at Sary Shagan along with BMD components in the early 1960s.

The USSR decided to develop SA-5A Surface to Air Missiles (SAM) against the B-70 bomber under development at that time in the US.

1960   Khrushchev promulgated a new military doctrine, "A Revolution in Military Affairs", consisting of reliance of nuclear missiles at the expense of strategic bombers and conventional forces.

April 1960   U.S. U-2 overflights were detected by Soviet prototypes of BMD radars and interceptors at Sary Shagan. These radars were later incorporated into the Moscow BMD system, some elements of the "Leningrad system," as well as the prototype of Hen House early warning radars which were deployed on the Soviet borders during the 1960s. It became evident that a major Soviet BMD program was underway and that a considerable progress towards the development of BMD components had already occurred.

1961   Construction began at the "Leningrad system" sites. In the early 1960s the USSR deployed its first BMD components near Leningrad and Moscow; at the same period, it began Soviet laser research with BMD implications.

Oct. 1961   The USSR conducted a series of atmospheric nuclear tests at Sary Shagan, testing black-out effects on the BMD and Early Warning radars under development there. Missiles were launched from the Kapustin Yar test range into the impact area in conjunction with BMD activities at Sary Shagan. The test series was repeated a year later. Taking the prospects of BMD deployment seriously, the USSR was concerned about dealing with the operational realities of BMD.

Defense Minister Malinovsky asserted that "the problem of destroying ballistic missiles in flight had been successfully solved." There was an increase of Soviet optimism and a surge of open Soviet commentary in the first half of the 1960s about the technical feasibility of an effective nation-wide defense against ballistic missiles.

Early 1960s   Deployment of the Soviet SA-5A SAMs at the "Tallin Line" of air defense installations (successor to the "Leningrad system"). Subsequently, 1960s SA-5A launchers were deployed at 130 sites in the USSR; plus, in the 1980s, some were deployed in Eastern Europe, Syria and Libya.

The PVO began to test Moscow BMD system components at Sary Shagan.

1962   Initial deployment of the two-stage, Griffon missile at 30 launch positions within the Leningrad system began.

July 1962   Khrushchev boasted that Soviet BMD interceptors could "hit a fly in outer space." In 1962-63, the USSR was claiming the lead in the field of BMD.

Sept. 1962   In a speech to the UN general Assembly, Soviet Foreign Minister A. Gromyko lamented Mutual Assured Destruction (MAD) and favored a new regime featuring a mutual build-down of offensive forces with an "exception for a limited number of anti-missile missiles and anti-aircraft missiles." He claimed to support a "general and complete disarmament." Thus, both Soviet military doctrine and "disarmament doctrine" supported BMD and other forms of active defense. Yet they diverged on the issue of strategic offensive forces. The military, recognizing the defensive-offensive relationship, stressed sharp mutual cuts of offensive arms. On the whole, however, arms control had a limited place in Soviet considerations regarding basic force levels at that time.

Late 1962   US intelligence detected preparation of BMD sites around Moscow. By that time a decision had been made by the Soviet leadership to produce and deploy the Moscow BMD system.

Jan. 1963   Commander of the Soviet PVO, Marshall Biryuzov, confirmed that "the problem of destroying enemy missiles in flight has been successfully solved. We have developed and constructed the means for defense of the country from a nuclear missile attack by an opponent."

Early 1963   Actual deployment of the Moscow BMD system began. Because of their lack of the desired capabilities in BMD and high-altitude air defense, the installations at the "Leningrad system" sites were dismantled, while the work on the Tallin air defense line proceeded.

Nov. 1963   The first public display of the Griffon missile was made at the Red Square parade.

Late 1963   Construction activity at the Moscow BMD sites became sporadic, indicating a technical problem.

1964   A special new section of the PVO, titled the PKO, "Anti-Space Defense," was established.

Oct. 1964   An article by Major General N. Talensky "Anti-Missile Systems and Problems of Disarmament," stressed:
                 - rejection of MAD as a destabilizing doctrine, making a state's security depending on a goodwill of potential adversary.
                 - acknowledgement of a dialectic interaction of offense and defense in arms development.
                 - the defensive character of Soviet BMD.
                 - maintenance of a "harmonious combination" of strong offenses and effective defenses (including the necessity to create a Soviet BMD system) as a major way to increase strategic stability.

Nov. 1964   The Galosh missile, the interceptor of the Moscow BMD system, was paraded through the Red Square.

Early 1966   General Secretary L. Brezhnev authorized full-scale ABM sites construction around Moscow.
                 The USSR started to deploy the SS-9 ICBM. This signified the beginning of the Soviet strategic offensive forces build-up. Also, a decision to produce and deploy the SA-5A SAM was made by the Soviet leadership, constituting a major upgrade of the Soviet air defense capabilities. Thus, taking into account the above mentioned Moscow BMD deployment, the USSR was proceeding in the mid-1960s with a combined offensive and strategic defensive forces, contrary to MAD doctrine.

Marshal P. Batitsky took charge of PVO. Advancement of the concept of "air-space defense" included: expanded and integrated air defense, BMD and anti- satellite cooperation under centralized control. In the mid-1960s the USSR considered defensive damage-limitation to be a major factor in successfully fighting a nuclear war and continued to intensify its efforts in strategic air defense, civil defense and BMD.

April 1966   Malinovsky announced that PVO could intercept all aircraft and "many" missiles, thus reaffirming (though less ambitiously) the success of Soviet BMD program.

1964-1967   The second period of activity at the Moscow BMD sites began, with the construction pace at its peak in 1966-1967. However, by 1967, the work went forward only on 6 of the 8 sites originally prepared for deployment.

Mid-1960s   The SA-2 SAM deployment began. At the same period, the USSR proceeded with full deployment of the Hen House early warning radars. A total of 11 radars were eventually deployed at 6 locations on the Soviet borders. In the mid-1960s, the USSR also began development of small BMD radars (Flat Twin and Pawn Shop) as well as of Soviet ASAT system. By the 1970s, the ASAT achieved an operational capability to intercept and destroy satellites in low orbits.

1967-1972: Reappraisal of Soviet BMD Policy

Jan. 1967   As a first reaction to the US proposal to negotiate the limits of ABM deployment renewed in Dec. 1966, the Soviet leadership stated that: because of the action-reaction interplay of strategic offensive and defensive forces, both must be included in such discussions. Thus both offensive and defensive systems should be limited, with the ultimate goal of "general and comprehensive disarmament." The USSR resisted US enthusiasm for limiting arms control to defensive systems.

1967   President Lyndon Johnson decided to deploy a light ABM Defense, the Sentinel. It was technologically more advanced than the Soviet BMD systems.

Intense debate in the USSR took place with respect to doctrine, technical prospects, and the strategic implications of ABM deployment. PVO chief Marshal Batitsky and other senior air defense officers, pushing for the construction of large ABM systems, claimed the PVO could meet the technical challenge of BMD and "reliably" protect the USSR against missile attack. Yet, leaders of the Strategic Rocket forces, Marshal N. Krylov, Civil Defense, Marshal V. Chuikov, and the Minister of Defense, Marshal A. Grechko, as well as some other Soviet senior military officials argued that some missiles would inevitably penetrate Soviet defenses. They implied that, at the prevailing level of military technological development, the best way to counter U.S. strategic power was through further expansion of Soviet strategic offensive forces and increasing their capabilities to provide damage denial in case of a major crisis. The PVO spokesman, Major-General Zavialov, counter-argued that the difficulty of destroying all enemy ballistic missiles through preemptive attack meant that the Soviet ABM systems had a vital role to play in achieving a favorable correlation of strategic forces during wartime.

In the first half of 1968 the debate continued on the same lines. On the whole, doubts about existing Soviet technology and the effectiveness of a nation-wide BMD, led to a more skeptical treatment of Soviet BMD from outside the PVO in the second half of 1960s. It also resulted in a general reappraisal of the Soviet policy towards BMD by the end of the decade.

This latter fact was also related to a Soviet doctrinal shift, first expressed in 1967 by the Commander-in-Chief of Soviet Strategic Missile Forces, Marshal Krylov. He pressed for a switch to the "Launch-on-Warning" concept as the primary measure of damage-limitation for strategic nuclear forces. This approach, however, required extensive early warning capabilities.

Among other things, this resulted in a political decision not to expand the Soviet BMD system beyond the Moscow region and to decrease the number of planned BMD installations around the Soviet capital.

The signing of the US-Soviet Treaty, banning weapons of mass destruction in outer space, took place.

Feb. 1967   Soviet Prime Minister, N. Kosygin, stated at a London press conference that BMD systems were not a factor in the arms race and strategic instability, and that cost-effectiveness alone should not dissuade a state from acquiring strategic defensive system. Soviet leaders were generally undecided on the issue of BMD limitations in 1967, despite displaying some signs of willingness to discuss it. Moreover, at that time, the Soviet leadership disagreements over the desirability of deploying a BMD system were intertwined with a more general debate over the wisdom of entering Soviet-American talks on strategic arms. Senior military officials and some Party leaders challenged the Brezhnev/Gromyko assumptions that it was possible for the U.S. and the USSR to cooperate to slow the development of military technology.

Sept. 1967   An "unofficial" paper, presented at a Pugwash conference by Col.

General A. Gryzlov, General Staff Liaison with Soviet Foreign Ministry on disarmament matters, echoed Gen. Talensky's pro-BMD views, defending the morality of ABMs in contrast to ICBMs.

Dec. 1967　　An unofficial meeting of US and Soviet scientists interested in arms control and disarmament, came to be especially useful and significant in influencing Soviet thinking away from the "Talensky doctrine" on BMD and arms control toward the "McNamara Doctrine," Mutual Assured Destruction (MAD).

June 1968　　Soviet foreign Minister A. Gromyko announced that the USSR was prepared to discuss limitations and reductions of offensive and defensive strategic systems "including ABMs." The initial positive Soviet reply to the US offer to include ABM issues in the SALT talks was expressed in a letter from Kosygin to Johnson in Jan. 1968. However, during the first half of 1968 Soviet leadership was still undecided as to whether to accept the limits on BMD.

July 1968　　The Nuclear Non Proliferation Treaty was completed.

Mid-1968　　Deployment of the Moscow BMD system slowed down and construction stopped completely at two of the six sites due to both technical and political obstacles. The Soviet decision to negotiate on BMD limits reflected increasing skepticism about Soviet BMD technology, as well as consideration of cost effectiveness. At the same time, the Galosh ABM achieved initial operating status in the Fall of 1968. The US congress authorized funding to begin deployment of the Sentinel, the light American BMD system.

Nov. 1968　　The opening round of the SALT-I talks began. Explicit references to Soviet BMD systems and to the concept of "air-space defense" disappeared from Soviet military publications, and likewise ABMs were not shown in Soviet military parades. This signified that the Soviet leadership had restrictions about BMD.
More generally, by this time, the Soviet political leadership accepted arms control as an important element of political-military strategy and as means of meeting Soviet security interests. However, this approach was not universally liked in the Soviet military establishment. Therefore, Gromyko, in his July 1968 address, had to criticize unidentified "theoreticians" who "try to persuade us...that disarmament is an illusion."

Late 1960s　　The USSR began to build laser facilities at Sary Shagan to investigate the performance of ground-based laser BMD.

Late 1960s-   Articles appeared in the Soviet military literature, discussing the possi-
early 1970s   bility of using relatively small mobile radars to perform the functions
               of Large Phased Array Radars (LPARs).

1969   The Nixon administration decided to support a phased ABM deploy-
               ment program, the SAFEGUARD system, which embodied substan-
               tially more sophisticated technology than did the Moscow BMD sys-
               tem. More generally, by the end of 1960s, the conclusion of some
               Soviet economists that the USSR was less technologically dynamic
               than the West filtered into the outlook of Soviet foreign policy
               specialists and some Politburo members. This influenced both Soviet
               policy of BMD deployment and the USSR's positions at the SALT-I
               talks.

May 1969   A major article by General Zemskov, editor of the confidential General
               Staff theoretical journal "Military Thought," appeared. The basic
               points were: the "nuclear balance" had been established, but it could
               be disrupted either by a sharp change in offensive capabilities or by the
               deployment of one side of a highly effective BMD, while the other side
               lags considerably behind. Thus, the conceptual reorientation of the
               USSR's senior political leadership caused certain shifts in official
               Soviet military doctrine as well, with the latter accepting certain ele-
               ments of the "arms control" mentality.

Late 1969   The chief of the Soviet SALT-I delegation recognized that MAD was a
               strategically stabilizing factor and that there were potentially
               destabilizing effects of ABM systems. Some Soviet diplomats
               accepted that an effective ABM, when deployed, would create
               uncertainty about the ability of a victim to mount a retaliatory strike,
               thus conceivably generating a temptation for a offensive first strike.
               This could stimulate an offensive-defensive spiral in strategic
               weapons. Subsequently these views were disseminated among Soviet
               defense analysts, especially those from the CPSU Central Committee
               apparatus, Foreign Ministry and Academy of Sciences. This signified
               that the "Talensky Doctrine," favoring ABM active defenses, was
               superseded, at least among civilian strategists, by concern over the
               "arms race stability" achievable through a controllable ABM/offensive
               missile interaction.
               The MAD logic as such, however, was not explicitly espoused by
               Soviet strategists, especially among the military, partly because of
               ideological considerations. It was impossible, at that time, for Soviet
               analysts to admit openly the possibility that the socialist USSR might
               attack first. It was also partly because of the influence of the Soviet
               military doctrine, aimed at survival in and winning a nuclear war, if

one occurred. Some Soviet defense analysts regarded comprehensive, balanced, and gradual disarmament as a better way to strengthen national security of the USSR.

By this time, the Soviet leadership had decided to seek the maximum ABM limitation consistent with maintaining a minimum Moscow defense against third country attack and accidental nuclear launches.

1970   Due to difficulties in negotiating strategic offensive arms limitations, the Soviet leadership decided to reach agreement first on the ABM Treaty and only then to return to offensive arms limitations.

The Moscow BMD system becomes operational. The main components were:
- Dog House and Cat House LPARs;
- 4 BMD complexes composed of two identical installations each comprising 3 mechanically steered Try Add radars (one for target acquisition, one for control and one to track the RV for ground-zero prediction) and 8 above-ground launchers for Galosh missiles (a total of 64 launchers); and
- A computer center for data management and processing.

A Soviet SA-5A ABM with a nuclear warhead was deployed.

1970-1971   The Soviet leadership decided to upgrade its radar network with Pechora-type LPARs. During the 1970s a total of 8 such radars were built on the Soviet borders.

1971   First appearance of the Flat Twin (phased array, target acquisition) and Pawn Shop (mobile, mechanically steered, interceptor guidance) small, rapidly deployable radars at Sary Shagan (a total of 6 radars built). During the 1970s and 1980s Flat Twin and Pawn Shop radars were operated at the sites where the new Galosh and Gazelle missiles were undergoing testing.

Late 1971   A Soviet decision was made to suspend the tests of its ASAT weapon.

Early 1970s   A ruby laser facility was deployed at Sary Shagan. At the same period, the Soviet research began to explore the technical feasibility of a particle beam weapon in space. The USSR was making the most of the provisions of the ABM Treaty, allowing BMD research.

Politburo members P. Shelest and A. Shelepin resisted the shift toward Detente with the West and, particularly, the conclusion of strategic arms control agreements with the U.S. This generated a conflict within the Communist Party leadership. In the mid-1970s, however, under the pressure of Brezhnev and his supporters, the policy consensus within the Party elite was restored. Shelest and Shelepin were removed from the Politburo in 1973 and 1975 respectively, while Grechko's death in 1976 paved the way for D. Ustinov, Brezhnev's longtime political ally to become Defense Minister.

May 1972   The signing of the ABM Treaty took place during U.S.-Soviet summit in Moscow. The Treaty permitted each side to deploy two 100-launcher BMD systems (one at the national capital and one to protect ICBMs), and to conduct further BMD research (but not to proceed with development).

1972-1983: Compliance with the ABM Treaty While Pushing of BMD Research and Incremental Modernization of Permitted Existing BMD Systems up to the Limits of the Treaty

June 1972   An article by two chief Soviet SALT-I negotiators stressed the interconnection of offensive and defensive arms development producing the arms race. They argue that, only by limitation on the buildup of both offensive and defensive weapons could the arms race be controlled. Soon afterwards, the moderating effect of the BMD limitations on the pace of weapons innovation was confirmed by Politburo member, N. Podgorny, Deputy Foreign Minister, Kuznetsov, Minister of Defense, A. Grechko, and chief of the General Staff, N. Kulikov. This became the official Soviet rationale for stating its ABM Treaty compliance.

Sept. 1972   Discussion and ratification of the ABM Treaty by the Supreme Soviet took place. Defense Minister, Marshal A. Grechko, while acknowledging that the Treaty would have a strategically stabilizing effect, emphasized that it "does not impose any limits whatsoever on research and experimental projects directed toward solving the problem of defending the country from nuclear rocket strikes."
Grechko and the senior PVO officers (the latter being deeply upset with the ABM Treaty on budgetary grounds) wished to maintain the Soviet program of ABM research. They believed that the problem of BMD would ultimately find an effective technical solution. They were especially concerned with the possibility that the US, which was substantially ahead in BMD technology, would find such a solution first. So Grechko's statement foreshadowed the later internal struggle over the pace and scope of Soviet ABM research.

1973   PVO chief Marshal Batitsky, in an article in the Soviet General Staff publication "Military Thought" revived the term "air-space defense." He stated that, because of the ABM Treaty restrictions, BMD would most probably change only qualitatively and would be limited in potential, capable of defending only a nation's capital. Thus, the BMD mission had not entirely disappeared from the PVO agenda in the 1970s, and a minority view continued to favor an enhanced BMD effort after the ABM Treaty was signed.

Moreover, on occasion these views were supported by some senior non-PVO military officials, revealing the dissatisfaction on the part of a significant segment of Soviet military establishment, especially within the General Staff. There was concern with the limitations posed by the ABM Treaty and the arms control mentality. In 1973, for example, the chief of Soviet General Staff Marshal V. Kulikov in an article in a PVO journal called to "ensure the protection of the country and armed forces from air and nuclear ballistic missile attack."

Nevertheless, although such proposals may have helped fuel the USSR's substantial program of BMD research, the notion of actually building a large BMD system had not been accepted by the bulk of the Party leadership, and also appeared to be undesirable for several military services competing with the PVO for institutional prominence and budgetary funds.

The USSR tested the SA-5 SAM radar in an ABM mode at Sary Shagan. The decision, in all probability, was taken at the level of PVO command supervising testing at Sary Shagan. Such radar tests soon stopped after high-level intervention by the Soviet political leadership, caused by the negative US reaction.

In the first half of the 1970s, there was an accelerated build-up of Soviet offensive strategic forces. Most Soviet strategists attributed to offensive nuclear weapons both the functions of damage denial (preemptive strike) and retaliation (second strike). They saw the ABM Treaty as means of protecting against third country or accidental nuclear strikes. This, however, did not preclude the option of using of the Moscow BMD system to protect the Soviet capital against US retaliation (diminished by Soviet preemptive strike). The BMD was seen as one of many measures like air defense and passive protection from civil defense.

1974   The signing of a Protocol to the US-Soviet ABM Treaty limited each side to a single, 100-launcher, ABM site. This could be used either for protection of the national capital or for ICBM fields. The USSR, having adopted as early as 1967 the Launch-on-Warning concept, never completely accepted MAD, and showed no real interest in BMD for the defense of its strategic forces. BMD was maintained to defend Moscow.

1975   The US decision to deactivate the newly-built SAFEGUARD BMD
              system at Grand Forks was seen as a favorable move in Moscow.

              The PVO dismantled a Flat Twin radar at Sary Shagan, transported it
              to Kamchatka and reassembled it in a few months, increasing US con-
              cerns about the possibility of creation of Soviet nation-wide BMD
              based on mobile components.

              Despite the protests of the Soviet Defense Minister, A. Grechko, the
              CPSU General Secretary, L. Brezhnev, decided to decrease the growth
              of the Soviet defense budget and to shift economic priorities towards
              consumer goods production vis-a-vis the military establishment and
              heavy industry. The PVO's budget became threatened among others.

Mid 1975   In a book by V. Bondarenko, *Modern Science and the Military
              Development*, published by the Soviet Defense Ministry, several
              points were made:
              - Detente is not irreversible and MAD is not a guarantee of Soviet
                security because of a) the West's irrational hatred of socialism, and
                b) the possibility of the emergence of new defensive technologies
                abrogating deterrence;
              - The impossibility of controlling the military technology by political
                means dictated the need to constantly modernize Soviet military
                technology (and at present - to accelerate these efforts) to prevent
                Western superiority in this field;
              - The dialectic of offensive and defensive weapons required accelerated
                development of the both types of armaments in the interests of Soviet
                national security.
              Roughly at the same time a book, *The Development of Air Defense*, by
              the Marshal of Aviation, G. Zimin, was published. Its major point was
              that counter-force strikes were not sufficient to protect the USSR in
              case of nuclear war; and, therefore, strategic air defense, including
              BMD, was needed to ensure military success.
              Later on, under pressure of the Soviet political leadership, this doc-
              trinal challenge, expressing the views of a significant group of the
              Soviet military elite and some high Party officials. It failed to affect the
              basic Soviet policy toward strategic defense. However, it contributed to
              acceleration of Soviet R&D efforts in the fields of BMD and ASAT
              weapons. For instance, the tests of the latter were resumed unilaterally
              the same year (1976) after a five- year break.

Jan. 1977   L. Brezhnev, gave a speech in Tula, attempting to discredit the aim of
              victory in a nuclear war and disavowed nuclear superiority as a Soviet
              goal. However, throughout the second half of the 1970s the USSR

was undertaking an intensive military build-up, emphasizing offensive might. In 1977, for instance, the decision was made to deploy highly accurate SS-18 ICBMs, considerably enhancing the USSR's first (or preemptive) strike capabilities. The USSR was reluctant to let the Treaty interfere with air defense, early warning, research in exotic technologies and other strategic defense efforts.

1977   The signing took place of the first U.S.-Soviet Common Understanding, prohibiting "concurrent testing" of air defense and BMD components under the terms of the ABM Treaty.

Late 1970s   The PVO expressed concerns that modern US cruise missiles posed a new threat to the USSR and argued for an increase in the anti-missiles capability of the Soviet air defense. Roughly at the same time Soviet cruise missile designer, V. Chelomey, proposed to Brezhnev that the USSR pursue a space-based kinetic energy BMD system. The proposal was rejected, however, by a special commission headed by Deputy Defense Minister for Armaments, V. Shabanov. The perspective of a nation-wide BMD system had little appeal in the Soviet political leadership at that time.

July 1978   BMD proponent, Marshal Batitsky, was demoted from the post of PVO commander. The PVO budget began to shrink in absolute terms along with the budget of Soviet Strategic Rocket Forces.

Mid-1978   The USSR entered negotiations with the US to ban ASAT weapons and suspended its own tests of its ASAT. Among other reasons, the Soviet government was not committed to a rapid upgrading of the ASAT system it had already created or to the rapid development of a more effective alternative utilizing sophisticated infra-red sensors.
             However, in the 1980s, after the US declined to continue ASAT talks in response to Soviet invasion of Afghanistan, the USSR resumed its ASAT testing and reported the development of a larger booster enabling Soviet ASAT to attack US satellites above a height of 15,000 miles in geo-synchronous orbits.

1980   The USSR began to deploy SA-10 SAM, allegedly having some limited ABM capabilities.

             Some Soviet observers, and especially PVO spokesmen, began to express apprehension that the shifting US outlook, exemplified by the non-ratification of the SALT-II agreement, might lead the US to discard the ABM Treaty and to create a nation-wide BMD system.

Removal began of half of the 64 Galosh launchers at the Moscow ABM sites, which signified the beginning of the upgrade program of the Moscow BMD system. Simultaneously, an active program of Soviet BMD research was pushed. The Soviet leaders, thus, renewed their commitment to a limited BMD deployment within the terms of the ABM Treaty. The major motivation was a desire to improve protection against attacks from the minor nuclear powers, as well as against a accidental launches. The late 1970s were the years of deterioration of Sino-Soviet relations due to sharp friction over Soviet aggression in Afghanistan, the Chinese attack against Vietnam, and the Cambodian problem.

At the same period the Soviet / West European detente was deteriorating, and the prospects of US INF deployment in Europe increased. In the Soviet view, this raised the possibility of a major nuclear conflict of the continent. The INF problems in Europe resulted in the USSR's decision taken at the same time to develop its own Anti-Tactical Ballistic Missile (ATBM).

March 1981   In a speech at the 26th CPSU Congress, Brezhnev called for the negotiation of superpower restraint on the creation of "qualitatively new" types of weapons.

Early 1980s   A CO_2 laser facility appeared at Sary Shagan.

Aug. 1981   The Soviet delegation to the UN tabled a draft of a multilateral treaty banning the militarization of outer space. The draft did not prohibit air-launched and ground-based ASAT weapons. Ten months later, in June 1982, the USSR again suspended ASAT tests and declared a testing moratorium as part of a new bid for ASAT talks.

1981   A major reorganization and renaming of the PVO took place, eliminating the title "national" form Air Defense Troops. There was a transition of control over tactical air defense units to Ground Forces though this was reversed in 1986. The transfer of 45% of PVO aircraft to the Soviet Air Force and sizable cuts in PVO budget, took place as well.

Early-to-   Backed by high-level party leaders including G. Romanov and V.
mid 1980s   Scherbitsky, elements of the Soviet military high command, most prominently, Deputy Defense Minister Marshal N. Ogarkov, raised an open challenge to established party policy. They opposed countering the US's new military and political assertiveness by a patient strategy of political maneuver, public diplomacy and substantive negotiations. For instance, in 1982, Ogarkov claimed that, because of the US new

assertiveness in the military sphere, the danger of war was increasing and the Soviet-American detente could not be reestablished on acceptable terms. This left no alternative except prolonged political hostility and an all-out arms race with the US. Therefore, he called, for a sharp acceleration of the development and deployment of Soviet weaponry and for a significant increase in Soviet military spending.

In response, Soviet Foreign Minister and Politburo member, A. Gromyko, offered a justification of detente, clearly directed at "domestic critics." He expressed the views of the Foreign Ministry and the bulk of the party elite (including Brezhnev and Chernenko) in an article, published in the main Party journal *Kommunist* in Sept. 1982. Gromyko urged relying on diplomacy and not on new weaponry to counter the US military build-up and political offensive. (Similar articles also appeared in *Kommunist* in Sept. 1984, signifying the continuation of the debate, then centered on the desirability of renewal of the Geneva talks with the U.S.).

During a period of growing U.S. assertiveness and Soviet foreign policy set- backs, Brezhnev partly yielded to the pressure of the military, and, in Oct. 1982, promised to accelerate Soviet military R&D, but refused to increase procurement.

1983-1985: Active Anti-SDI Efforts

March 1983 President Reagan gave his SDI speech on March 23.

General Secretary Y. Andropov argued in an article that SDI would undercut the basic U.S.-Soviet understanding about the interaction between offensive an defensive weapons and result in undermining strategic stability. This article initiated a major Soviet anti-SDI campaign, which was characterized by an increased Soviet attention to the ABM Treaty, emphasis on its absolute necessity as a pivotal factor of strategic stability, and by the claim that the U.S. was exceeding the Treaty terms with SDI. The SDI was seen as part of a major US drive to achieve strategic superiority and overturn the international position of parity previously attained by the USSR.

1983 Soviet delegation broke off the Geneva talks in protest against the beginning of the US INF deployments in Europe.

Construction of a Pechora-type LPAR started in Abalakovo, near Krasnoyarsk. The planning for the Krasnoyarsk radar was begun in 1980-81 at the Soviet Ministry of Defense. This undertaking was a clear violation of the ABM Treaty.

1983-1984   Field tests of Soviet SA-X-12B "Giant" SAM against the medium-range SS-12 missile were begun. The low success rate, one hit for twenty missiles, meant that this SAM needed to be significantly improved before it could serve as an effective ATBM. The less advanced SA-12A "Gladiator" version was being deployed throughout the 1980s with the Soviet Ground Forces for tactical air defense.

Aug. 1983   As a part of the anti-SDI campaign, the Soviet delegation to the UN tabled a draft of a multilateral treaty prohibiting the use of force in outer space and from space to earth. This included a ban on creation and testing of ASAT systems. In the same month, Andropov announced a Soviet moratorium on the placement of Soviet ASAT weapons in space, which would be observed as long as the US refrained from tests of its ASAT system.

Early 1984   Marshal Ogarkov warned against the West's stepped-up development of sophisticated weapons based on "new physical principles" and asserted that a Soviet counter-program must immediately be launched. Ogarkov emphasized the Western military threat inherent in Follow-On Force Attack (FOFA) technologies. He did not refer directly to SDI and did not favor a commitment to deploy an extensive Soviet BMD system.
             In May 1984, in a major article in the main Soviet Defense Ministry newspaper, *Red Star*, Ogarkov reiterated his claim, warning that the failure to respond promptly to the challenge of accelerating Western military programs would be a serious error which would undermine Soviet security.

April 1984   The publication of the report by "The Working Group of Soviet Scientists for Peace and Against the Nuclear Threat," the Sagdeyev Report, contained a detailed survey of possible Soviet military countermeasures against the SDI. However, the applicability and ultimate effectiveness of the proposed measures was questionable and was largely drawn from Western sources.

June 1984   After a short period of internal political struggle with Politburo members Romanov and Scherbitsky and backed by majority of the Politburo members (including M. Gorbachev), Chernenko offered to the US a resumption of the Geneva talks without the precondition of NATO INF rollbacks. The specific terms of these talks were agreed on between the U.S. and the Soviet Union in Jan. 1985. The talks got underway in March 1985.
             At the opening round of the Geneva Defense and Space talks, the Soviet delegation insisted on the "narrow" interpretation of the ABM

Treaty and called for a total ban on SDI research. The initial Soviet position, linking progress in START and INF to US agreement to such ban, was supported by the Soviet military elite. The military leadership's views were expressed in a June 1985 article in *Pravda*, by the new chief of Soviet General Staff, Marshal Akhromeyev.

Sept. 1984 General Secretary K. Chernenko, emphasizing economic needs and diplomatic means in dealing with the West, refused to support a larger military effort. After his unconditional proposal to the US to resume the Geneva talks, Chernenko demoted Ogarkov. This made the hardliners more cautious, but did not eliminate the debate itself.

Early 1985 Soviet Foreign Minister A. Gromyko informed his West German counterpart that the Kremlin would view the Bonn government as an "accomplice" in violating the ABM treaty if it helped the U.S. with the SDI. Similar pressure was put on France during Gorbachev's visit to Paris in the fall of 1985. This part of the Soviet anti-SDI campaign was designed to erode US allies' support for the program.

May 1985 The Soviet Defense Minister, Marshal Sokolov, speaking on SDI, warned that the USSR would not be driven down any military investment path laid out for it by the U.S. He confirmed that any SDI deployment would lead to both defensive and offensive countermeasures by the USSR.

Also, the director of the largest Soviet center for laser and nuclear fusion research, N. Basov, asserted that the USSR would have "no scientific problem in developing lasers capable of intercepting missiles in space." The initial Soviet reaction to SDI included an effort to make an impression that the USSR could easily offset SDI militarily.

1985-1990: The Gorbachev Era

1985 At a meeting with senior Soviet officers, General Secretary M. Gorbachev, called for stringent limits on military expenditures and for channeling more resources toward industrial modernization. Economic considerations then became one of the major factors providing for Soviet policy aimed at preserving the ABM Treaty restraints.

The USSR proposed to halt construction at the Krasnoyarsk LPAR if the US would stop improvements at Thule and Fylingdales early warning sites in Greenland.

June 1985 Gorbachev forced Romanov, his leading conservative rival, out of the Politburo.

A year later, the CPSU Secretary responsible for overseeing foreign policy, A. Dobrynin, acknowledged that Gorbachev's heightened focus on diplomacy and compromises with the US had been accompanied by "fierce collisions, sharp discussions and painful disagreements" within the Soviet political elite. Some of these disagreements have centered on how to cope with SDI and, more generally, what policy to pursue concerning Strategic Defense.

The signing of a second U.S.-Soviet Common understanding prohibited the parties from switching an air defense component at a test site where BMD components are operating.

Sept. 1985   In an interview granted to *Time* magazine, Gorbachev appealed to American public opinion, in an effort to erode popular support for the SDI program.

Oct. 1985   The Soviet delegation in Geneva put forward a proposal for a 50% reduction in strategic weapons, predicated on the prohibition on all SDI research. This proposal was unsuccessfully pushed by Gorbachev during his Geneva summit with President Reagan in Nov. 1985 and at Reykjavik summit in Oct. 1986.

Late 1985   Shortly after the Gorbachev/Reagan Geneva summit, Politburo member Scherbitsky and the Chief of General Staff Marshal Akhromeyev gave a skeptical account of the summit results. They expressed doubts about the possibility of reaching common understanding with the US through negotiations, and claimed that the preservation of peace hinged on increasing USSR's military might and praised the "firmness" of the Soviet negotiating approach. Subsequently, Gorbachev muted this implicit challenge to his political line and emphasized diplomatic, rather than military means to counter the US military build-up and to derail the American SDI.

Jan. 1986   Gorbachev put forward a plan for nuclear disarmament by the year 2000, according to which the US and the USSR would forswear BMD and reduce strategic offensive arms in a series of cuts. This initiative also comprised a proposal to decouple the SDI and INF issues. Thus led to the signing of the INF Treaty two years later. The Soviet military elite implicitly doubted the value of these policies.

Feb. 1986   The second version of the "Sagdeyev Report" listed new suggested Soviet military countermeasures against SDI.

Spring 1986   At the 27th CPSU Congress, Gorbachev proclaimed the principle of

"Reasonable Sufficiency" as a guideline for the Soviet defense build-up. This was the onset of the "New Thinking" in the Soviet military policy, which linked military issues to political and economic ones.

May 1986   The weakening of the linkage between SDI, START and INF became clear to the Soviet delegation attending the talks in Geneva. The Soviets accepted the SDI-type research in the laboratory, but wanted tighter terms of the ABM Treaty to restrict the testing of exotic ABM systems and proposed a US-Soviet agreement on not withdrawing from the ABM Treaty for a minimum of 15 years.

Mid-1986   A drastic shake-up of the Soviet diplomatic corps took place. This strengthened Gorbachev's ally, CPSU Secretary Dobrynin, and his control of Soviet foreign policy-making.

Late-1986   Western experts were invited to examine the Flat Twin and Pawn Shop radars at Sary Shagan. After this inspection took place, the USSR complied with the Western requests; and, in the late 1986 and early 1987, dismantled these radars. This signified emergence of a new, cooperative Soviet approach towards treaty-compliance disputes. In 1988 the components of a radar installation in Gomel (Belorussia) were completely destroyed on the request of the U.S. inspectors, who had visited this installation in December 1987.

Deployment of the mobile version of the Soviet SA-10 SAM began. By 1990, this missile constituted 25% of the USSR's 9,000 strategic SAM launchers, with 1,850 of SA-10 being deployed near installations of strategic importance.

Marshal Akhromeyev emphasized that the ABM Treaty permitted the testing of a land-based ABM system, based on "other physical principles" and able to protect one region of each country. It was hoped such a Soviet system could be a major step toward an effective defense-suppression weapon for use against a US space-based BMD system.

Nov. 1986   At the US-Soviet Reykjavik summit, Gorbachev stated that the USSR would find a relatively inexpensive "asymmetrical" response to SDI, implying Soviet reliance primarily on offensive and defense-suppressing weapons to counter SDI. However, part of the Soviet top military, including Marshal Akhromeyev and some conservative Party leaders like Scherbitsky, called for creation of the USSR's own extensive BMD system along with augmenting offensive weapons if the U.S. moved to SDI systems deployment. In their view, the Soviet program of economic revitalization could be sacrificed for this end.

1987   The USSR called a halt to construction on the Krasnoyarsk radar, and invited a group of U.S. congressmen to visit it. The Soviet government announced that the radar was being turned over to the USSR Academy of Science and suggested that the Krasnoyarsk facility could be used as an international station for space exploration and research.

May 1987   At a Warsaw Pact summit, a new "defensive doctrine," implying both defensive military strategy and defensive force structure, was adopted. However, in the field of strategic arms, the emphasis remained on the dominance of offensive weapons to the detriment of strategic defense.

Late 1980s The Committee on Defense and State Security of the Supreme Soviet was created for legislative supervision of Soviet military programs. The general trend towards greater civilian involvement in military affairs provided for further strengthening of the political approach to military issues and restraining the hard-liners among top military elite.

June 1988   The CPSU Conference declared that Soviet military development would take a qualitative, not a quantitative course, stressing technological sophistication despite declining military budgets. This was welcomed by the Soviet military elite, pressing its technological agenda. At the same time this signified that the Soviet leadership became more pessimistic about USSR's technological capabilities vis-a-vis the West than it was twenty years before.

July 1988   At a Soviet Foreign Ministry Conference, Foreign Minister Shevardnadze emphasized the political side of security policy, argued for greater civilian involvement in military affairs, and demanded that decisions on military development pass through Foreign Ministry for assessment of their compliance with arms control agreements.

1989   A group of U.S. congressmen and representatives of the U.S. National Defense Resources Council were invited to inspect the Soviet CO_2 laser facility at Sary Shagan. The group stated that Soviet CO_2 lasers at Sary Shagan were low-power, used to track aircraft and satellites.

            A modernized Moscow BMD system became operational. The main components were:
            - The integrated Pill Box LPAR;
            - 100 reloadable launchers in hardened silos with 2 types of interceptors;
            - An exoatmospheric, modernized version of Galosh, and an endoatmospheric missile, named Gazelle;
            - Advanced computer center for data management and processing.

The USSR signalled its willingness to conclude a START accord despite the lack of agreement on strategic defense. Gorbachev's line of Soviet foreign policy-making prevailed over the military hardliners.

Summer 1989
The former chief of Soviet SALT-II delegation, V. Karpov, told a U.S. Congressional group visiting Moscow, that the USSR might consider dismantling the Moscow BMD system in exchange for dismantling of the decommissioned SAFEGUARD at Grand Forks. Soviet analysts began to question in public the value of the Moscow BMD system on the grounds of its technical and cost-effectiveness as well as its implication for strategic stability.

Sept. 1989
The Soviet decision to dismantle the Krasnoyarsk LPAR as violating the ABM Treaty was done despite complaints of the Soviet military that this would harm the USSR's security. The violation was publicly acknowledged by Soviet Foreign Minister Shevardnaze in his speech to the Supreme Soviet a month later, in October 1989.

April 1990
A report from the U.S. intelligence community stated that there had been no signs of cuts in the Soviet spending on strategic defense, despite reductions elsewhere in the Soviet defense budget.

May-June 1990
At the U.S.-Soviet summit in Washington, the two sides committed themselves to beginning another round of strategic arms control negotiations soon after completion of the START I Treaty. One of the goals was "to implement an appropriate relation between strategic offenses and defenses."

July-Aug. 1990
Several articles by Soviet military and civilian experts, proposing U.S.-Soviet cooperation in development of a ground- and space-based BMD for protection against accidental launches and limited attacks by third countries were published. This reflected growing Soviet concern about the proliferation of nuclear weapons and ballistic missiles in the Third World.

Appendix C

Chemical and Biological Weapons

TABLE A   Effects of the Chemical Sarin

Weather Conditions[a]	Agent Form[b]	Area Effected (hectares/ton)	
		Lethal Dose	Incapacitating Dose
Clear,	vapor	10-18	26-37
sunny day,	a/v	17-32	47-64
light breeze	aerosol	29-48	73-94
Overcast with	vapor	9-16	13-34
moderate wind,	a/v	19-37	27-83
day or night	aerosol	37-63	54-150
Clear,	vapor	150-460	420-1300
calm night	a/v	260-400	590-820
	aerosol	240-430	380-900

Note: Assumes LCt_{50} of 70 mg.min/m^3 and ICt_{50} of 35 mg.min/m^3 (appropriate for mildly active, unprotected men), for releases of 100 to 1,000 kilograms of Sarin on an urban target.

a.   "clear, sunny day, light breeze" corresponds to Pasquill class "A" stability for residential urban areas, a mixing height of 2,000 meters, and a wind speed of 2 meters per second; "overcast" corresponds to class "D" stability, a mixing height of 1,000 meters, and a wind speed of 5 meters per second; "clear, calm night" corresponds to class "F" stability, a mixing height of 250 meters, and a wind speed of 1 meter per second.

b.   a/v = 50 percent fine aerosol, 50 percent vapor. The velocity is assumed to be 0.01 meters per second for a fine aerosol, 0 meters per second for vapor.

Source: S. Fetter, "Ballistic Missiles and Weapons of Mass Destruction: What is the Threat? What Should be Done?" *International Security*, Summer 1991, Vol. 16, No. 1, pp. 21.

TABLE B   Effects of Various Biological Agents

Pathogen	Disease	Respiratory ECt$_{50}$[a] (mg.min/m^3)	Time to Effect (days)	Mortality Rate (percent)
F. tularensis	Tularemia	0.001	2-5	0-60
B. anthracis	Anthrax	0.1	1-4	95-100
P. pestis	Plague	b	3-4	90-100
C. burnetii	Q fever	0.001	18-21	1-4
Vee virus	VEE	0.001	2-5	0-2

a.   Median pathogen dosage that would produce the disease in resting, unprotected men, assuming agent-infested particles with diameters of 1 to 5 microns, and assuming that a fraction of the organisms die during dissemination (95 percent for Tularemia, 50 percent for Anthrax, 90 percent for Q fever, and 80 percent for VEE).

b.   Plague is highly contagious; the number of people exposed to a given concentration of the pasteurella pestis bacteria would not be an accurate indication of how many people would eventually contract the disease. A dose of about 3,000 bacteria per man would result in a 50 percent probability of contracting the disease.

Source: Stockholm International Peace Research Institute, *The problem of Chemical and Biological Warfare, Vol II: CB Weapons Today* (New York Humanities Press, 1973), pp. 42-43, as cited in S. Fetter, "Ballistic Missiles and Weapons of Mass Destruction: What is the Threat? What Should be Done?" *International Security,* Summer 1991, Vol. 16, No. 1, pp. 25.

Appendix D

Section of President Reagan's March 23, 1983 Speech Concerning Strategic Defense

...Now, thus far tonight I've shared with you my thoughts on the problems of national security we must face together. My predecessors in the Oval Office have appeared before you on other occasions to describe the threat posed by Soviet power and have proposed steps to address that threat. But since the advent of nuclear weapons, those steps have been increasingly directed toward deterrence of aggression through the promise of retaliation.

This approach to stability through offensive threat has worked. We and our allies have succeeded in preventing nuclear war for more than three decades. In recent months, however, my advisers, including in particular the Joint Chiefs of Staff, have underscored the necessity to break out of a future that relies solely on offensive retaliation for our security.

Over the course of these discussions, I've become more and more deeply convinced that the human spirit must be capable of rising above dealing with other nations and human beings by threatening their existence. Feeling this way, I believe we must thoroughly examine every opportunity for reducing tensions and for introducing greater stability into the strategic calculus on both sides.

One of the most important contributions we can make is, of course, to lower the level of all arms, and particularly nuclear arms. We're engaged right now in several negotiations with the Soviet Union to bring about a mutual reduction of weapons. I will report to you a week from tomorrow my thoughts on that score. But let me just say, I'm totally committed to this course.

If the Soviet Union will join with us in our effort to achieve major arms reduction, we will have succeeded in stabilizing the nuclear balance. Nevertheless, it will still be necessary to rely on the specter of retaliation, on mutual threat. And that's a sad commentary on the human condition. Wouldn't it be better to save lives than to avenge them? Are we not capable of demonstrating our peaceful intentions by applying all our abilities and or ingenuity to achieving a truly lasting stability? I think we are. Indeed, we must.

After careful consultation with my advisers, including the Joint Chiefs of Staffs, I believe there is a way. Let me share with you a vision of the future which offers hope. It is that we embark on a program to counter the awesome Soviet missile threat with measures that are defensive. Let us turn to the very strengths in technology that spawned our great industrial base and that have given us the quality of life we enjoy today.

What if free people could live secure in the knowledge that their security did not rest upon the threat of instant U.S. retaliation to deter a Soviet attack, that we could intercept and destroy strategic ballistic missiles before they reached our own soil or that of our allies?

I know this is a formidable, technical task, one that may not be accomplished before the end of this century. Yet, current technology has attained a level of sophistication where it's reasonable for us to begin this effort. It will take years, probably decades of effort on many fronts. There will be failures and setbacks, just as there will be successes and breakthroughs. And as we proceed, we must remain constant in preserving the nuclear deterrent and maintaining a solid capability for flexible response. But isn't it worth every investment necessary to free the world from the threat of nuclear war? We know it is.

In the meantime, we will continue of pursue real reductions in nuclear arms, negotiating from a position of strength that can be ensured only by modernizing our strategic forces. At the same time, we must take steps to reduce the risk of a conventional military conflict escalating to nuclear war by improving our non-nuclear capabilities.

America does possess--now--the technologies to attain very significant improvements in the effectiveness of our conventional, non--nuclear forces. Proceeding boldly with these new technologies, we can significantly reduce any incentive that the Soviet Union may have to threaten attack against the United States or its allies.

As we pursue our goal of defensive technologies, we recognize that our allies rely upon our strategic offensive power to deter attacks against them. Their vital interests and ours are inextricably linked. Their safety and ours are one. And no change in technology can or will alter that reality. We must and shall continue to honor our commitments.

I clearly recognize that defensive systems have limitations and raise certain problems and ambiguities. If paired with offensive systems, they can be viewed as fostering an aggressive policy, and no one wants that. But with these considerations firmly in mind, I call upon the scientific community in our country, those who gave us nuclear weapons impotent and obsolete.

Tonight, consistent with our obligations of the ABM treaty and recognizing the need for closer consultation with our allies, I'm taking an important first step. I am directing a comprehensive and intensive effort to define a long-term research and development program to begin to achieve our ultimate goal of eliminating the threat posed by strategic nuclear missiles. This could pave the way for arms control measures to eliminate the weapons themselves. We seek neither military superiority nor political advantage. Our only purpose--one all people share--is to search for ways to reduce the danger of nuclear war.

My fellow Americans, tonight we're launching an effort which holds the promise of changing the course of human history. There will be risks, and results take time. But I believe we can do it. As we cross this threshold, I ask for your prayers and your support.

Thank you, good night, and God bless you.

Bibliography

Acheson, D., Testimony of the Secretary of State, Hearings of the Senate Foreign Relations Committee, *Assignment of Ground Forces of the United States in the European Area*, February, 1951.

Adams, B. *Ballistic Missile Defense*, New York: American Elsevier Publishing Co., 1971.

Adelman, K., *The Great Universal Embrace: Arms Control - a Skeptic's Account*, New York: Simon and Schuster, 1989.

Anderson, M. *Revolution - The Reagan Legacy*, Stanford, Cal.: Hoover Institution Press, 1990.

Apple, R., "Ukraine Gives in on Surrendering Its Nuclear Arms," *New York Times*, January 11, 1994, p. A-1.

Arbatov, A., "Russia's Foreign Policy Alternatives," *International Security*, Fall 1993, Vol. 18, No. 2, pp. 5-43.

Armstrong, S., "Patriot Missile Successes in Gulf Give 'Star Wars' New Credibility," *Christian Science Monitor*, February 6, 1991.

Art, R., "A U.S. Military Strategy for the 1990s: Reassurance Without Dominance," *Survival*, Winter 1992-92, Vol 43, No. 4, pp. 3-22.

Aspin, L., "The Bottom-Up Review: Forces for a New Era," *Mimeo.*, Washington, D.C.: U.S. Department of Defense, September 1, 1993.

_____ and W. Dickinson, *Defense for a New Era - Lessons of the Persian Gulf War*, Washington, D.C.: Brassey's 1992.

_____ . "A New Kind of Threat - Nuclear Weapons in an Uncertain Soviet Union," Washington, D.C.: House Armed Services Committee, September 12, 1991.

Aviation Week and Space Technology, "Pentagon Science Advisers Criticize SDIO's Rush to Adopt Brilliant Pebbles," April 9, 1990, p. 23.

Awanohara, S., "My Shield or Yours?" *Far Eastern Economic Review*, October 14, 1993, p. 22.

Bailey, M., "The Market Mechanism in the Defense Department," in R. McKean, ed., *Issues in Defense Economics*, New York: Columbia University Press for the N.B.E.R., 1967.

Bailey, K. and R. Rudney, eds., *Proliferation and Export Controls*, Lanham, Md.: University Press of America, 1993.

Bailey, K., *Strengthening Nuclear Non-Proliferation*, Boulder, Colo.: Westview Press, 1993.

Ballistic Missile Defense Organization, *1993 Report to the Congress on the Theater Missile Defense Initiative*, Washington, D.C.: U.S. Department of Defense, 1993.

Barfield, C. and W. Schambra, eds., *The Politics of Industrial Policy*, Washington, D.C.: A Conference Sponsored by the American Enterprise Institute for Public Policy Research, 1986.

Baruch, B. "The Baruch Plan," Statement by the U.S. Representative to the United Nations Atomic Energy Commission, June 14, 1946, in U.S. Department of State, *Documents of Disarmament, 1945-1959, Vol.1*, Washington, D.C.: U.S. Government Printing Office, 1960, p. 7.

Baucom, D. *The Origins of SDI 1944-83*, Lawrence, Kansas: University of Kansas Press, 1992.

Betermier, Adm. J., et. al., *Proliferation and Missile Defenses*, Fairfax, Va.: Study Group on Proliferation and Missile Defense, NIPP, June 1993.

Bethe, H., R. Garwin, K. Gottfried, and H. Kendall, "Space Based Ballistic Missile Defense," *Scientific American*, October 1984, Vol. 251, No.4, pp. 39-44.

Bialer, S., and J. Afferica, "Reagan and Russia," *Foreign Affairs*, Winter 1982/1983, Vol. 61, No. 2, pp. 249-71.

Blacker C. and C. Duffy, eds., *International Arms Control - Issues and Agreements*, Stanford, CA: Stanford University Press, 1984.

Blair, B. and H. Kendall, "Accidental Nuclear War," *Scientific American*, December 1990, Vol. 263, No. 6.

Blechman, B. and V. Utgoff, "The Macroeconomics of Strategic Defenses, " *International Security*, Winter 1986-87, Vol. 11:3, pp. 33-70.

Bloembergen, N. and C.K.N. Patel et al., "Report to the APS of the Study Group on Science and Technology of Effected Energy Weapons," Executive Summary and Major Conclusions, *Physics Today*, May 1987, pp. S3-S16.

Bobrow, D., ed., *Weapons System Decisions: Political and Psychological Perspectives on Continental Defense*, New York: Frederick A. Praeger Publishers, 1969.

Boffey, P., "Obstacles Force Narrower Focus on 'Star Wars'," *New York Times*, October 19, 1986, p. 1.

Boffey, P.M., W. Broad, L. Gelb, C. Mohr, and H. Noble, *Claiming the Heavens - Complete Guide to the Star Wars Debate*, New York: Times Books, 1988.

Borrus, M., "Reversing Attrition: A Strategic Response to the Erosion of U.S. Leadership in Microelectronics," *BRIE*, Working Paper No. 13, Berkeley, Cal.: University of California, March 1985.

Bracken, P., *The Command and Control of Nuclear Weapons*, New Haven: Yale University Press, 1983.

_____ . "Nuclear Weapons and State Survival in North Korea," *Survival*, Autumn 1993, Vol. 35, No. 3, p. 137.

Brams, S. "Theory of Moves," *American Scientist*, Nov.-Dec. 1993, Vol. 81, pp. 562-570.

_____ and D.M. Kilgour, "Deterrence Versus Defense: A Game-Theoretic Model of Star Wars," *Mimeo.*, 1987.

Brennan, D., ed., *Arms Control, Disarmament, and National Security*, New York: Brazilier, 1961.

Broad, W., "Pentagon Starts Project to Judge Anti-Missile Plan," *New York Times*, August 6, 1987, p. 1.

_____ . "The Secrets of Soviet Star Wars," *New York Times Magazine*, June 28, 1987, p. 22.

_____ . "Space Station Is Studied by Military," *New York Times*, April 7, 1987, p. C-1.

_____ . "Congress preparing to Quietly Approve an Antimissile Plan," *New York Times*, November 18, 1991, p. A-1.

_____ . "Pentagon Analyst Questions Timing for Missile Shield - Crippling Problems Seen," *New York Times*, June 2, 1992, p. A-1.

_____ . *Teller's War - The Top Secret Story Behind the Star Wars Deception*, New York: Simon & Schuster, 1992.

Brodie, B., *Sea Power in the Machine Age*, Princeton, N.J.: Princeton University Press, 1941.

_____ . *The Absolute Weapon*, New York: Harcourt Brace, 1946.

Brown, H., "Is SDI Technically Feasible?" *Foreign Affairs - America and the World 1985*, Vol. 64, No. 3, pp. 435-54.

Bundy, McG. et al., "Reykjavik's Grounds for Hope," *New York Times*, October 19, 1986, p. E-23.

_____ . "The Bishops and the Bomb," *The New York Review of Books*, June 16, 1983.

_____ , G. Kennan, R. McNamara, and G. Smith, "The President's Choice: Star Wars or Arms Control," *Foreign Affairs*, Winter 1984/1985, Vol. 63, No. 2, pp.264-78.

Calaway, P.R., *European Cooperation in Research and Development*, Washington, D.C.: National Defense University Press, 1989.

Canby, S. and I. Dorfer, "More Troops, Fewer Missiles," *Foreign Policy*, Winter 1983/84, No. 53, pp. 3-17.

Caravelli, J. "Soviet Countermeasures to SDI," *Journal of Defense Diplomacy*, March 1985, pp. 45-47.

Carlucci, F., *1989 Report to the Congress on the Strategic Defense Initiative*, Washington, D.C.: U.S. Department of Defense, S.D.I.O., March 13, 1989.

Carnesale, A., "Managing the ABM Treaty Regime: Issues and Options," in *Defending Deterrence - Managing the ABM Treaty into the 21st Century*, eds., Antonia Chayes and P. Doty, Washington, D.C.: Pergamon-Brassey's, 1989.

Carnesale, A., and R. Haass, eds., *Superpower Arms Control - Setting the Record Straight*, Cambridge, MA: Ballinger, 1987.

Carter, A., and D. Schwartz, eds. *Ballistic Missile Defense*, Washington, D.C.: Brookings Institution, 1984.

Carus, W.S. and J. Bermudez, "Iraq's Al-Husayn Missile Program," *Jane's Soviet Intelligence Review*, June 1990.

Carus, W.S., *Ballistic Missiles in Modern Conflict*, New York: Praeger, 1991.

Charles, D., "Rise and Fall of Star Wars," *The New Scientist*, March 20, 1993.

Chayes A. and J. Weisner, eds., *ABM: An Evaluation of the Decision to Deploy an ABM System*, New York, Harper and Row, 1969.

Chayes, Antonia and P. Doty, eds., *Defending Deterrence - Managing the ABM Treaty into the 21st Century*, Washington, D.C.: Pergamon-Brassey's, 1989.

Cheney, R., *Report of the Secretary of Defense to the President and the Congress*, Washington, D.C.: U.S.G.P.O., January 1990 & 1991.

Clapper, Lt. Gen. J., Jr., "Remarks" at the Senate Select Intelligence Committee Hearing, January 25, 1994, Washington, D.C.: Federal News Service, January 25, 1994.

Codevilla, A., "Space, Intelligence, and Deception," *Mimeo.*, presented to a panel at the U.S. Naval Postgraduate School, September 1985.

_____. "Who Killed SDI? We could have had a working anti-missile defense 30 years ago. We still don't. Why?" *National Review*, May 10, 1993, pp. 40-43.

Cohen, D., "The Non-Report That Would Never Die," *USC Information Sciences Institute*, April 1987.

_____ et al., "A Report to the Director of SDIO," Eastport Study Group, *Mimeo.*, December 1985.

Cohen, E., "The Mystique of Air Power," *Foreign Affairs*, January/February 1994, Vol. 73, No. 1, pp. 109-24.

Congressional Record, 91st Congress, 1st session, July 9, 1969, 115: 18910.

Cooper, H., "End of Tour Report," *Mimeo.*, Washington, D.C.: U.S. Department of Defense, January 20, 1993.

_____ and S. Hadley, "Briefing on the Refocused Strategic Defense Initiative edited Transcript," *Mimeo.*, Washington, D.C.: U.S. Department of Defense, February 12, 1991.

Cushman, J. "Pentagon Official Proposes Cost Cut for Space Weapon," *New York Times*, September 8, 1988, p. A-1.

_____. "Senate Endorses Pact to Reduce Strategic Arms," *New York Times*, October 2, 1992, p. A-6.

Dannreuther, R. "Russia, Central Asia, and the Persian Gulf," *Survival*, Winter 1993-94, Vol. 35, No. 4.

Davis, W., Jr., *Asymmetries in U.S. and Soviet Strategic Defense Programs: Implications for Near-Term American Deployment Options*, Washington, D.C.: Pergamon-Brassey's International Defense Publishers, 1986.

Defense Intelligence Agency, *Soviet Military Space Doctrine*, Washington, D.C.: D.I.A. #DDB-1400-16-84, August 1, 1984.

Denoon, D., "The SDI: Dilemma for Allied Cooperation," a paper presented at the AEI/Adenaur Stiftung Conference on German-US Relations, April 1986.

_____. *Real Reciprocity - Balancing U.S. Economic and Security Policies in the Pacific Basin*, New York: Council on Foreign Relations, 1993.

DiMaggio, C. and R. Civiak, "The Strategic Defense Initiative: A Model for Estimating Launch Costs," *Congressional Research Service*, CRS Report for Congress, p. 87-475 SPR, June 4, 1987.

_____ et al., "The Strategic Defense Initiative: Program Description and Major Issues," *Congressional Research Service*, Report no. 86-8 SPR, January 7, 1986.

Doder, D., "Soviets See U.S. 'Deception'," *Washington Post*, January 7, 1985, p. A-1.

Drell, S., P. Farley, and D. Holloway, *The Reagan Strategic Defense Initiative: A Technical, Political, and Arms Control Assessment*, Stanford, Cal.: Stanford Center for International Security and Arms Control, 1984.

Dulles, J.F., "Policy for Security and Peace," *Foreign Affairs*, April 1954, Vol. XXXII, No. 3.

Durch, W., "Protecting the Homeland," in *The American Military in the 21st Century*, B. Blechman, et. al., New York: St. Martin's Press, 1993., pp. 226-230.

Dyson, F. "A Case for Missile Defense, " *Bulletin of the Atomic Scientists*, April 1969, Vol. 25.

The Economist, "Japan Wants Star Without Wars," August 2, 1986, p. 69.

The Emerging Ballistic Missile Threat to the United States, Report of the Proliferation Study Team, Washington, D.C.: U.S. Department of Defense, February 1993.

Engelberg, S., and M. Gordon, "Intelligence Study Says North Korea Has Nuclear Bomb," *New York Times*, December 26, 1993, p. A-1.

Enthoven, A., and K.W. Smith, *How Much Is Enough? Shaping the Defense Program 1961-1969*, New York: Harper & Row, 1971.

Fairbanks, C. Jr. and A. Shulsky, "From 'Arms Control' to Arms Reductions: The Historical Experience," *Washington Quarterly*, Summer 1987, Vol. 10, pp. 59-73.

Fetter, S., "Ballistic Missiles and Weapons of Mass Destruction: What is the Threat? What Should be Done?" *International Security*, Summer 1991, Vol. 16, No. 1, p. 27.

Field, G. and D. Spergel, "Cost of Space-Based Laser Ballistic Missile Defense," *Science*, March 21, 1986, Vol. 231, pp. 1387-93.

Finletter, T., *Power and Policy*, New York: Harcourt Brace, 1954, p. 392.

Fletcher, J. et al., *The Strategic Defense Initiative - Defensive Technologies Study*, Washington, D.C.: US Department of Defense, April 1984.

Fossedal, G., "Let's Get the Facts on the Arms Talks," *New York Times*, February 11, 1987, p. 23.

Fraumeni, B., "Investment in Education and U.S. Economic Growth," *Mimeo.*, Cambridge, Mass.: 1991.

Freedman, L., *The Evolution of Nuclear Strategy*, New York: St. Martin's Press, 1989.

Friedman, T., "U.S. Vision of Foreign Policy Reversed," *New York Times*, September 22, 1993, p. A-13.

_____ . "U.S. Formally Rejects 'Star Wars'" in ABM Treaty," *New York Times*, July 15, 1993, p. A-6.

_____ . "U.S. - Russia Accord on Arms Hits Snag," *New York Times*, October 15, 1992, p. A-3.

_____ . "U.S. Offers to Negotiate on 'Star Wars'," *New York Times*, October 16, 1991, p. A-3.

Fukuyama, F., "The End of History," *The National Interest*, Summer 1989, No. 16, pp. 3-18.

Gaddis, J.L., "International Relations Theory and the End of the Cold War," *International Security*, Winter 1992/1993, Vol. 17, No. 3, pp. 5-58.

Gaffney, F. Jr., "Hapless SDI Damsel in Distress," *The Washington Times*, June 10, 1991, p. D-3.

Garwin, R., "The Soviet Response," in *Empty Promise - The Growing Case Against Star Wars*, ed., J. Tirman, Boston: Beacon Press, 1986, pp. 129-146.

_____ . "Are Brilliant Pebbles All That Brilliant?" *Aerospace America*, December 1990, p. 6.

Gejdenson, S. and T. Roth, "Export Policies for the 1990s," *New York Times*, July 18, 1994, p. A-15.

General Board, "V-2 Rocket Attacks and Defense," Document 502.101-42 in Air Force Historical Research Center, Maxwell Air Force Base, Alabama.

Gilmartin, T., "Early SDI System May Cost $60 Billion," *Defense News*, March 23, 1987, Vol. 2, No. 12, p. 1.

Gleick, J., "New Superconductors Offer Chance to Do the Impossible," *New York Times*, April 9, 1987, p. A-1.

Goldfischer, D., *The Best Defense*, Ithaca, N.Y.: Cornell University Press, 1993.

Gordon, M., "Pentagon Panel Calls for Testing of 6 Programs for Missile Shield," *New York Times*, August 7, 1987, p. A-11.

_____ . "U.S. Hopes to Curb A-Arms by Restricting Fuel Output," *New York Times*, July 28, 1993.

_____ . "Military Plan would Cut Forces But Have Them Ready for 2 Wars - Clinton Strategy Endorses Bush's Basic Doctrine," *New York Times* September 2, 1993, p. A-1.

_____ . "U.S. Seeking to Loosen Missile Defense Curbs," *New York Times*, December 3, 1993, p. A-14.

_____ . "North Korea Test Cruise Missile Designed to Sink Ships," *New York Times*, June 1, 1994, p. A-12.

_____ . "North Korea Said to Have A-Bomb Fuel - Pace of Program Alarms Washington," *New York Times*, June 8, 1994, p. A-7.

Gray, C., *Nuclear Strategy and National Style*, Lanham, Md.: Hamilton Books, 1986.

_____ . *House of Cards - Why Arms Control Must Fail*, Ithaca, NY: Cornell University Press, 1992.

Graybeal, S., "Testimony before the Senate Foreign Relations Committee on Effective Theater Missile Defenses and the ABM Treaty," *Mimeo.*, May 3, 1994.

_____ and P. McFate, *The ABM Treaty and Ballistic Missile Defense*, Washington, D.C.: AAAS Publication #93-26s, 1993.

Guerrier, S.W. and W.C. Thompson, *Perspectives on Strategic Defense*, Boulder, Colo.: Westview Press, 1987.

Haass, R. and A. Carnesale, "Arms Control Plus 'Star Wars'," *New York Times*, February 2, 1987, p. A-21.

Haberman, C., "The Middle East Poker Game - After Assad's Play, the Pressure is Now on Israel," *New York Times*, January 17, 1994, p. A-16.

Hall, C., *Britain, America and Arms Control, 1921-1937*, New York: St. Martin's Press, 1987.

Halperin, M., "The Decision to Deploy the ABM: Bureaucratic and Domestic Politics in the Johnson Administration," *World Politics*, October 1972, Vol. 25.

Hao, Y. and G. Huan, *The Chinese View of the World*, New York: Pantheon Books, 1989.

Hardin, R., J. Mearsheimer, G. Dworkin, and R. Goodin, eds., *Nuclear Deterrence - Ethics and Strategy*, Chicago, Ill: University of Chicago Press, 1985.

Hecht, J., *Beam Weapons: The Next Arms Race*, New York/London: Plenum Press, 1984.

Hersh, S., "Missile Wars," *The New Yorker*, September 26, 1994, pp. 90-94.

Hitch, C. and R. McKean, *The Economics of Defense in the Nuclear Age*, Cambridge, Mass.: Harvard University Press, 1967.

Hitch, C., *Decision Making for Defense*, Berkeley, Cal.: University of California Press, 1965.

Hoffert, M., et al., "Earth-to-Satellite Microwave Beams: An Innovative Approach to Onboard Space Power Requirements," *Mimeo.*, August 1987.

Hoffert, M. and G. Miller, "Ballistic Missile Defense: Cost of Space-Based Laser," Rebuttal to Field and Spergel, March 1986, *Science*, November 28, 1986, Vol. 234, pp. 1056-59.

Hoffman, D. and J. Young, "A Hallway Huddle Caps 9 Years of Arms Bargaining," *Washington Post National Weekly Edition*, July 22-28, p. 16.

Hoffman, F. et al., eds, *Swords and Shields: NATO, the USSR, and New Choices for Long-Range Offense and Defense*, Lexington, Mass.: Lexington Books, 1987.

_____ . *Ballistic Missile Defense and U.S. National Security*, Washington, D.C.: U.S. Department of Defense, 1983.

Holdren, J. and F.B. Green, "Military Spending, the SDI, and Government Support of Research and Development: Effects on the Economy and the Health of American Science," *Journal of the Federation of American Scientists FAS*, F.A.S. Public Interest Report, September 1986, Vol. 39, No. 7.

Holmes, S., "China Denies Violating Pact by Selling Arms to Pakistan," *New York Times*, July 26, 1993, p. A-2.

_____ . "U.S. Determines China Violated Pact on Missiles," *New York Times*, August 25, 1993, p. A-1.

House, K., "Korea, Clinton's Cuban Missile Crisis," *Wall Street Journal*, January 5, 1994, p. A-15.

Huntington, S., "Conventional Deterrence and Conventional Retaliation in Europe," *International Security*, Winter 1983/84, Vol. 8, No. 3.

Ikle, C.F., "The Reagan Defense Program: A Focus on the Strategic Imperatives," *Strategic Review*, Spring 1982.

_____ . "Nuclear Strategy - Can There Be a Happy Ending?" *Foreign Affairs*, Spring 1985, Vol. 63, No. 4.

Ingersoll, B., "Patriot Missile's Success Against Iraq's Scuds Knocks Star Wars Prospects Into Higher Orbit," *Wall Street Journal*, January 29, 1991, p. A-20.

Inoguchi, T., "Japan's Response to the Gulf Crisis: An Analytic Overview," *Journal of Japanese Studies*, 1991, Vol, 17, No. 2, pp. 257-273.

International Study Group on Proliferation and Missile Defense, *Proliferation and Missile Defense: European-Allied and Israeli Perspectives*, Fairfax, Va.: National Institute for Public Policy, June 1993.

Interavia Space Directory, 1992-93.

Jarrell, R. and M. Cagel, *History of the Plato Anti-Missile Missile Program: 1952-60*, Redstone Arsenal, 1961.

Jastrow, R. and M. Kampelman, "How to Meet the Third World Missile Threat," *Wall Street Journal*, November 19, 1993, p. A-14.

Jehl, D., "U.S. Outlines Concern Over North Korean A-Arms," *New York Times*, February 25, 1993, p. A-7.

Jenkins, F., "Implications of Defenses and the ABM Treaty of U.S. Strategic Arms Control Policy in the 1990s," *Mimeo.*, Mclean, Va.: S.A.I.C., October 1991.

Jervis, R., "Arms Control, Stability, and Causes of War," *Political Science Quarterly*, 1993, Vol. 108, No. 2.

Jervis, R., N. Lebow and J. Stein, *Psychology and Deterrence*, Baltimore, Md.: The Johns Hopkins University Press, 1989.

Johnston, B. and W. Proxmire, "SDI's Broken Promises," *Wall Street Journal*, August 28, 1987.

Johnson, C., "MITI, NPT, and the Telecom Wars: How Japan Makes Policy for High Technology," *BRIE*, Working Paper No. 21, Berkeley, Cal.: University of California, September 1986.

Jorgensen, D. and B. Fraumeni, "Investment in Education and U.S. Economic Growth," *Mimeo.*, Cambridge, Mass.: Harvard University, 1991.

Kamien, M. and N. Schwartz, "Market Structure and Innovation: A Survey," *The Journal of Economic Literature*, March 1975, Vol. XIII, No. 1, pp. 1-37.

Kaplan, J., *Wizards of Armageddon*, New York: Simon & Schuster, 1983.

Kaufman, W., "The Crisis in Military Affairs," *World Politics*, July 1958, Vol. X, No. 4.

_____ . *The McNamara Strategy*, New York: Harper and Row, 1964.

_____ and J. Steinbruner, *Decisions for Defense - Prospects for a New Order*, Washington, D.C.: The Brookings Institution, 1991.

Kaysen, C., R. McNamara, and G. Rathjens, "Nuclear Weapons After the Cold War," *Foreign Affairs*, Fall 1991, Vol. 70, No. 4, p. 100.

Keeny, S., "Inventing An Enemy," *New York Times*, June 18, 1994, p. 21.

Kennan, G., "The Source of Soviet Conduct," *Foreign Affairs*, July 1947, Vol. XXV, No. 4. [X] is the name used by G. George for this article.

Kendrick, J., *The Formation of Stocks of Total Capital*, New York: National Bureau of Economic Research, 1976.

Kincade, W., "The SDI and Arms Control," in *Strategic Defenses and Soviet-American Relations*, eds., S. Wells and R. Litwak, Cambridge, Mass.: Ballinger, 1987.

Kissinger, H., *Nuclear Weapons and Foreign Policy*, New York: Harper, 1957.

Kolodziej, E., *Making and Marketing Arms - The French Experience*, Princeton, N.J.: Princeton University Press, 1987.

Kristof, N., "Clinton Aide Ends Trip With No Sign of Accord," *New York Times*, May 13, 1993, p. A-10.

Kugler, R., *Commitment of Purpose: How Alliance Partnership Won the Cold War*, Santa Monica, Cal.: Rand, 1993.

Kuhn, T. *The Structure of Scientific Revolutions*, 2nd. ed., Chicago, Ill.: University of Chicago Press, 1970.

Kurth, J., "Why We Buy The Weapons We Do," *Foreign Policy*, Summer 1973, No. 11.

Lakoff, S. and H. York, *A Shield in Space? Technology, Politics and the Strategic Defense Initiative*, Berkeley, Cal.: University of California Press, 1989, pp. 9- 15.

Lawrence Livermore National Laboratory, "Kinetic-Kill Vehicles," *Energy and Technology Review*, July 1987, pp. 16-17.

_____ . "Induction Accelerator and Free-Electron Laser Development, 1986-1987," *LLL TB 86*.

Lewis, J. and H. Di, "China's Ballistic Missile Programs: Technologies, Strategies, Goals," *International Security*, Fall 1992, Vol. 17, No. 2, pp. 5-40.

Lewis, P., "Ex-Foes Trade Stories From the Cold War Trenches," *New York Times*, March 1, 1993, p. A-7.

Lunn, S., *Burden-Sharing in NATO*, Lindon: Routledge and Kegan Paul, 1983.

Lustick, I., "Reinventing Jerusalem," *Foreign Policy*, Winter 1993-94, No. 93, pp. 41-59.

MacDonald, B., "Falling Star: SDI's Troubled Seventh Year," *Arms Control Today*, September 1991.

McDougall, W., *The Heavens and the Earth*, New York: Basic Books, Inc., 1985.

McKean, R., ed., *Issues in Defense Economics*, New York: National Bureau of Economic Research, 1967.

McNamara, R., "Defense Arrangements of the North Atlantic Community," *Department of State Bulletin*, July 9, 1962, Vol. 47.

March, J. and H. Simon, *Organizations*, New York: John Wiley, 1959.

_____ . Statement before a Joint Session of the Senate Armed Services Committee and the Senate Subcommittee on Department of Defense Appropriations on the Fiscal year 1968-72 Defense Program and the 1968 Defense Budget, January 23, 1967.

Mathias, C., Jr., "Why Did Gorbachev Shift on Missiles?," *New York Times*, April 2, 1987, p. A-31.

Mearsheimer, J., "Back to the Future: Instability in Europe After the Cold War," *International Security*, Summer 1990, Vol. 15, No. 1, pp. 5-56.

Milholin, G. and G. White, "Proliferation in Disguise," *New York Times*, July 18, 1994, p. A-15.

Miller, S. and S. Van Evera eds., *The Star Wars Controversy*, Princeton, NJ: Princeton University Press, 1986.

Millis, W., *Arms and Men: A Study in American Military History* New York: Putnam and Sons, 1956.

Mohr, C., "Antimissile Plan Seeks Thousands of Space Weapons," *New York Times*, November 3, 1985, p. A-1.

Monahan, Lt. Gen. G., *Strategic Defense Initiative - FY 1991 Program Briefing*, Washington, D.C.: SDIO, February 2, 1990.

Nagler, R., *Ballistic Missile Proliferation - An Emerging Threat*, Arlington Va.: System Planning Corporation, 1992.

Nelson, R., ed.,*Government and Technical Progress*, New York: Pergamon Press, 1982.

_____. *High-Technology Polices: A Five Nation Comparison*, Washington and London: American Enterprise Institute for Public Policy Research, 1984.

New Republic, "Too Brilliant By Half," editorial, May 29, 1989, pp.7-9.

New York Times, editorial, "Mr. Nunn's Rash Rush to ABMs," July 29, 1991.

Newhouse, J., *Cold Dawn - The Story of SALT*, New York: Holt, Rinehart and Winston, 1973.

Nincic, M., "Can the U.S. Trust the U.S.S.R.?" *Scientific American*, April 1986, Vol. 254, pp. 33-41.

Nitze, P., "After Iraq, Nukes Can Be Junked," *Wall Street Journal*, December 24, 1991, p. 15.

_____. "SDI, Arms Control, and Stability: Toward a New Synthesis," *U.S. Department of State, Current Policy*, No. 845, Washington, D.C.: U.S. Government Printing Office, June 1986.

_____. *From Hiroshima to Glasnost*, New York: Grove, Weidenfeld, 1989.

Nolan, J., *Trappings of Power - Ballistic Missiles in the Third World*, Washington, D.C.: Brookings Institution, 1991.

_____. and A. Wheelon, "Third World Ballistic Missiles," *Scientific American*, August 1990.

_____. *Guardians of the Arsenal*, New York: New Republic Books, 1989.

Nollyn, R., *Personality and National Character*. New York: Pergamon Press, 1971.

Norman, C., "News and Comment," *Science*, January 16, 1987.

Norton, R.D., "Industrial Policy and American Renewal," *Journal of American Literature*, March 1986, Vol. XXIV, pp. 1-40.

Nunn, S., "Nunn Outlines New Arms Control and Strategic Modernization Agenda," *Mimeo.*, Washington, D.C.: Office of Senator Nunn, January 19, 1988.

Nye, J. and J. Schear, "SDI: A Lever We Should Use," *Washington Post*, December 15, 1986, p. A-15.

Odom, Lt. Gen. W., USA Ret., Chairman, The Proliferation Study Team, *The Emerging Ballistic Missile Threat to the United States*, Washington, D.C.: February 1993, p. 1.

Office of the Assistant Secretary of Defense for Public Affairs, "New Strategic Defense Initiative Focus: Global Protection Against Limited Strikes GPALS," *Mimeo.*, Washington, D.C.: Press release, January 30, 1991.

Office of the Secretary of Defense, *U.S./Russian Joint Statement on a Global Protection System*, *Mimeo.*, Washington, D.C.: June 16, 1992.

Osgood, R., *NATO, The Entangling Alliance*, Chicago, Ill: University of Chicago Press, 1962.

Osgood, R., *Limited War: The Challenge to American Strategy*, Chicago, Ill: University of Chicago Press, 1957.

Panel on National Security Controls in International Technology Transfer, *Balancing the National Interest*, Washington, D.C.: National Academy Press, 1986.

Panofsky, W., "The Strategic Defense Initiative: Perception vs. Reality," *Physics Today*, June 1985, pp. 34-45.

Payne, K., *Strategic Defense: "Star Wars" in Perspective*, Maryland and London: Hamilton Press, 1986.

_____ . *Missile Defense in the 21st Century: Protection Against Limited Threats*, Boulder, Colo.: Westview Press, 1991.

The Pentagon Papers, Gravel edition, Vol. 1, Boston, Mass.: Beacon Press, 1971, pp. 412-429.

Perlez, J., "Ukraine Hesitates on Nuclear Deal - Kiev Parliament Is Reserved on Plan to Yield Weapons," *New York Times*, January 12, 1994, p. A-1.

Perle, R., "Statement Before the Committee on Armed Services," U.S. House of Representatives, April 16, 1991.

Postol, T., "Lessons of the Gulf War Experience with Patriot," *International Security*, Winter 1991-1992, Vol. 16, No. 3, p. 141.

_____ . "Lessons from the Gulf War Patriot Experience: A Technical Perspective," *Mimeo.*, Testimony before the House Armed Services Committee, April 16, 1991.

Quade, E.S., ed., *Analysis for Military Decisions*, Amsterdam and London: North-Holland Publishing Company, 1970.

Quester, G., *Nuclear Diplomacy: The First Twenty-Five Years*, New York: Dunellen, 1970.

Radner, R., "Attrition, Deterrence, and the Value of Ballistic Missile Defense," *Mimeo.*, October 11, 1985, Murray Hill, NJ: Bell Laboratories, pp. 1-28.

Report of the President's Commission on Strategic Forces, Washington, D.C.: U.S. Government Printing Office, April 1983.

Report of the Secretary of Defense to the President and to the Congress, Washington, D.C.: U.S.G.P.O., 1990.

Reuters, "SDI, Chernobyl Are Said to Have Helped End Cold War," *Washington Post*, February 27, 1993, p. A-24.

Ricks, T., "How Wars Will Change Radically Says Pentagon Planner," *Wall Street Journal*, July 14, 1994, p. 1.

Rogers, B., "Why Compromise Our Deterrent Strength in Europe?," *New York Times*, June 28, 1987, p. E-17.

Rosenberg, N., "Civilian 'Spillovers' from Military R&D Spending: The American Experience Since World War II," *Mimeo.*, A paper presented at a Conference on Technical Cooperation and International Competitiveness, Lucca, Italy, April 1986.

_____ . *Inside the Black Box: Technology and Economic*, Cambridge, U.K.: Cambridge University Press, 1982.

Rosenthal, A., "U.S. Offers to Negotiate on 'Star Wars'," *New York Times*, October 16, 1991, p. A-3.

Rotberg, R., and K. Rabb, eds., *The Origin and Prevention of Major Wars*, Cambridge, U.K.: Cambridge University Press, 1989.

Rowny, E., "SDI: Enhancing Security and Stability," *Current Policy*, Washington, D.C.: U.S. Department of State, April 1988, No. 1058.

Safire, W., "China's Hama Rules," *New York Times*, March 5, 1992, p. A-27.

Sanger, D., "'Star Wars' Facing Cuts and Delays - '92 Goal in Doubt," *New York Times*, November 22, 1987, p. 40.

_____ . "War of Technology: Soviet Subs Run Silent," *International Herald Tribune*, June 13, 1987, p. 1.

_____ . "Many Doubtful of Early Move on 'Star Wars'," *New York Times*, February 11, 1987, p. 1.

_____ . "U.S.-North Korea Atom Accord Expected to Yield Dubious Results," *New York Times*, January 9, 1994, p. L-1.

_____ . "U.S. Presses Japan on Missile Project - Aspin Backs Away from Requiring Tokyo to Offer Technological Secrets," *New York Times*, November 13, 1993, p. A-12.

_____ . "Tokyo Raids Seek to Halt Aid for North Korea on Missiles," *New York Times*, January 15, 1994, p. L-5.

_____ . "North Korea Buying Old Russian Subs," *New York Times*, January 21, 1994, p. A-6.

Sapolsky, H. *The Polaris System Development: Bureaucratic and Programmatic Success in Government*, Cambridge, Mass.: Harvard University Press, 1972.

Sawyer, K., "Soviets Making Steady Gains in Space," *Washington Post*, July 20, 1987, p. A-1.

Schelling, T., *The Strategy of Conflict*, New York: Oxford University Press, 1960.

_____ and M. Halperin, *Strategy and Arms Control*, New York: 20th Century Fund, 1961.

Scherer, F., *International High Technology Competition*, Cambridge, Mass.: Harvard University Press, 1992.

Schmitt, E., "Agency Proposes Options to Cut 'Star Wars' Costs," *New York Times*, May 28, 1992, p. A-17.

SDI Monitor, "SDI Budget Summary Reveals Near-Term Tilt," April 22, 1987, p. 1.

_____ . "Effect of SDI Budget Cut Slowly Emerging," November 9, 1990, p. 249.

Segal, G., "The Coming Confrontation Between China and Japan," *World Policy Journal*, Summer 1993, Vol. X, No. 2, pp. 27-32.

Shapley, D., *Power and Promise*, Boston, Mass.: Little-Brown, 1993.

Shribman, D., "Kemp Joins GOP Race for Presidency, Vows to Make it Star Wars Referendum," *Wall Street Journal*, April 7, 1978, p. 70.

Shulsky, A., "From 'Arms control' to Arms Reductions: The Historical Experience," *Washington Quarterly*, Summer 1987, Vol. 10.

Shultz, G., *Turmoil and Triumph - My Years as Secretary of State*, New York: Charles Scribner's Sons, 1993.

Slocombe, W., "The Countervailing Strategy," *International Security*, Summer 1981, Vol. 6, No. 1.

Smith, R.J., "Schlesinger Attacks Star War Plans," *Science*, November 9, 1984, Vol. 226:4675, p. 673.

Spector, L., *Nuclear Ambitions - the Spread of Nuclear Weapons 1989-1990*, Boulder, Colo.: Westview Press, 1990.

Stein, R., "Patriot ATBM Experience in the Gulf War," *Mimeo.*, Lexington, Mass.: Raytheon Corp., Winter 1991.

_____ . "Correspondence: Patriot Experience in the Gulf War," *International Security*, Summer 1992, Vol. 17, No. 1, pp. 199-225.

Stowsky, J., "Beating Our Plowshares into Double-Edged Swords: the Impact of Pentagon Policies on the Commercialization of Advanced Technologies," *BRIE*, Working Paper No. 17, Berkeley, Cal.: University of California, April 1986.

The Strategic Defense Initiative Organization, "The New Focus for SDI: GPALS," *Mimeo.*, Washington, D.C.: U.S. Defense Department, June 6, 1991.

_____ . *SDI Briefing Charts*, Washington, D.C.: October 1988.

_____ . "Status of SDI Cost Initiatives and Analysis," May 1987.

_____ . "Civil Applications for Promising SDI Technologies," *Mimeo.*, Washington, D.C.: SDIO, 1987.

_____ . "Report to the Congress on the Strategic Defense Initiative Deployment Schedule," May 12, 1987.

_____ . "Report to the Congress on the Strategic Defense Initiative," April, 1987.

_____ . "Report to the Congress on the Strategic Defense Initiative," June 1986.

Strobel, W., "Plan Urges 'Star Wars' to Be Built Gradually," *Washington Times*, May 20, 1988, p. 4.

Talbott, S., *Master of the Game*, New York: Vintage Books, 1988,

Taylor, M., *The Uncertain Trumpet*, New York: Harper, 1960.

Thurow, L., "How to Wreck the Economy," *The New York Review of Books*, May 14, 1981.

Tirman, J., ed., *Empty Promise: The Growing Case Against Star Wars*, Boston: Beacon Press, 1986.

_____ . *The Fallacy of Star Wars*, New York: Vintage Books for the Union of Concerned Scientists, 1984.

Trainor, B., "A Missile-Free Europe: Little Impact on A War," *New York Times*, May 1, 1987, p. A-1.

Tucker, Robert, C., *Political Culture and Leadership in Soviet Russia from Lenin to Gorbachev*. New York: W.W. Norton and Company, 1987.

Tyler, P., "Chinese Military Sees U.S. As a Foe - Disaffection in Army Seen in Book Recalled By Beijing," *New York Times*, November 16, 1993, p. A-1.

_____ . "U.S. and China Agree to Expand Defense Links," *New York Times*, November 3, 1993, p. A-13.

U.S. Department of Defense, *Conduct of the Persian Gulf War, Final Report to the Congress*, Washington D.C.: Department of Defense, April 1992.

_____ . *Soviet Strategic Defense Programs*, Washington, D.C.: Department of Defense, October 1985.

_____ . *The Soviet Space Challenge*, Washington, D.C.: November 1987.

_____ . *Final Report to the Congress: Conduct of the Persian Gulf War*, Washington, D.C.: U.S.G.P.O., April 1992.

U.S. Executive Office of the President, Security Resources Panel of the Science Advisory Committee, *Deterrence and Survival in the Nuclear Age: Report to the President*, Washington, D.C.: November 7, 1957, p. 19.

Van Cleave, W., *Fortress USSR*, Stanford, Cal.: Hoover Institute Press, 1986.

Van Creveld, M., *Technology and War: From 2000 B.C. to the Present*, New York: Free Press, 1989.

Vlahos, M., *Strategic Defense and the American Ethos: Can the Nuclear World Be Changed?*, SAIS Papers, No. 13, Boulder, Colo.: Westview Press and FPI, 1986.

Waldman, P. and B. Schlender, "Is A Big Federal Role the Way to Revitalize Semiconductor Firms?," *The Wall Street Journal*, February 17,   1987, p. 1.

The Wall Street Journal, editorial, "Lawyers v. Defense," June 15, 1987.

_____ . "A Statistical Look at China - By the Numbers," December 10, 1993, p. R-12.

Wanner, B., "Washington, Tokyo Explore Defense Technology Cooperation," *JEI Report*, Washington, D.C.: Japan Economic Institute, October 8, 1993.

Warnke, P., "Apes on a Treadmill," *Foreign Policy*, Spring 1975, No. 18.

The Washington Post, "Defense Science Board Report on SDI," July 10, 1987, p. 21.

Weigley, R., "War and the Paradox of Technology," *International Security*, Fall 1989, Vol. 14, No. 2, p. 196.

Weinberger, C., and G. Shultz, *Soviet Strategic Defense Programs*, Washington, D.C.: U.S. Departments of State and Defense, October 1985.

_____ . "It's Time to Get SDI Off the Ground," *New York Times*, August 21, 1987, p. A-7.

_____ . "Strategic Defense in Perspective," *Defense - 1986*, January-February 1986, p. 6.

_____ . "Responding to Soviet Violations Policy Study," *Mimeo.*, Washington, D.C.: U.S. Department of Defense, November 13, 1985.

Weiner, T., "Lies and Rigged 'Star Wars' Test Fooled the Kremlin and Congress," *New York Times*, August 18, 1993, p. A-1.

_____ . "Patriot Missile's Success a Myth, Israeli Aides Say," *New York Times*, November 21, 1993, p. L-13.

_____ . "Inquiry Finds 'Star Wars' Tried Plan to Exaggerate Test Results," *New York Times*, July 23, 1994, p. 1.

Wells, S., Jr. and R. Litwak, ed., *Strategic Defenses and Soviet- American Relations*, Cambridge, Mass.: Ballinger Publishing Company, 1987.

Wiesner, J. "The Case Against an Antiballistic Missile System," *Look*, November 28, 1967.

Winnefeld, J. and J. Pollack et. al., *A New Strategy and Fewer Forces: the Pacific Dimension*, Santa Monica, Cal.: RAND, 1992.

Winograd, T., "Strategic Computing Research and the Universities," *Mimeo.*, Stanford: Stanford University, November 10, 1986.

Wohlstetter, A., "Between an Unfree World and None - Increasing Our Choices," *Foreign Affairs*, Summer 1985, Vol. 63, No. 5.

Wolfe, T., *Soviet Power and Europe, 1945-1970*, Baltimore, Md.: Johns Hopkins Press, 1970.

Wood, L. and G. Canavan, "Statement on American Physics Society Report Presented to the House Republican Research Committee," *Mimeo.*, May 19, 1987, pp. 1-10.

Woolsey, J., "World Trouble Spots," Hearing Before the Senate Select Intelligence Committee, Washington, D.C.: Federal News Service, January 25 1994.

Worden, S., *SDI and the Alternatives*, Washington, D.C.: National Defense University Press, 1991.

Yakushiji, T., "The Dynamics of Techno-Industrial Emulation," *BRIE*, Working Paper No. 15, Berkeley, Cal.: University of California, Summer 1985.

York, H. and J. Wiesner, "National Security and the Nuclear Test Ban," *Scientific American*, October 1964, pp. 27-35.

Yost, D., *Swords and Shields - NATO, The USSR and New Choices for Long-Range Offense and Defense*, Lexington, Mass.: Lexington Books, 1987.

Yost, D., *Soviet Ballistic Missile Defense and the Western Alliance*, Cambridge, Mass.: Harvard University Press, 1988.

Zakheim, D. and J. Ranney, "Matching Defense Strategies to Resources: Challenges for the Clinton Administration," *International Security*, Summer 1993, Vol. 18, No. 1, pp. 51-78.

Zegveld, W. and C. Enzing, *SDI and Industrial Technology Policy: Threat or Opportunity?*, New York: St. Martin's Press, 1987.

Zhang, L.Z., *People's Daily*, Overseas Edition, August 1, 1988, FBIS Daily Report China, August 4, 1988, pp. 41-42.

About the Book and Author

With the end of the Cold War and the visibility of U.S. Patriot missile defenses during the 1991 Gulf War, the cost and benefits of ballistic missile defense systems (BMD) need to be reevaluated. In this detailed and balanced study, David Denoon assesses new types of short-range and intercontinental missile defenses.

In the post-Cold War era, two fundamental changes have made missile defense for the United States and its military forces more compelling: The United States and Russia no longer see each other as direct threats, and there has been a dramatic proliferation of ballistic missile capability in the Third World. Consequently, U.S. forces deployed overseas are more likely to be at risk and, eventually, the United States itself could become vulnerable to missile threats.

With these changes in mind, David Denoon analyzes the current BMD dilemma, arguing that active defenses against missiles should be seen as a form of insurance against catastrophe. He assesses the likelihood of missile attacks and the appropriate level of investment for the United States to defend against such attacks. The book provides an assessment of deterrence and the performance of the Patriot missiles during the 1991 Gulf War, critiques the Strategic Defense Initiative, and analyzes the prospects for new types of short-range and intercontinental missile defenses.

David Denoon is a professor of politics and economics at New York University. He has a B.A. from Harvard University, an M.P.A. from Princeton University, and a Ph.D. from M.I.T. He was formerly vice president of the U.S. Export-Import Bank and Deputy Assistant Secretary of Defense and is the author of several books, including *Constraints on Strategy: The Economics of Western Security* and *Real Reciprocity: Balancing U.S. Economic and Security Policies in the Pacific Basin.*